Eagan Press Handbook Series

Starches

David J. Thomas and William A. Atwell

Cover: Pudding and corn dogs, ©1997 Artville, LLC;
Visco/amylo/Graph R courtesy of C. W. Brabender Instruments, Inc.;
spoon with gel courtesy of National Starch and Chemical Company.

Library of Congress Catalog Card Number: 98-71475
International Standard Book Number: 0-891127-01-2

Printed in the United States of America on acid-free paper

American Association of Cereal Chemists
3340 Pilot Knob Road
St. Paul, Minnesota 55121-2097, USA

About the Eagan Press Handbook Series

The Eagan Press Handbook series was developed for food industry practitioners. It offers a practical approach to understanding the basics of food ingredients, applications, and processes—whether the reader is a research chemist wanting practical information compiled in a single source or a purchasing agent trying to understand product specifications. The handbook series is designed to reach a broad readership; the books are not limited to a single product category but rather serve professionals in all segments of the food processing industry and their allied suppliers.

In developing this series, Eagan Press recognized the need to fill the gap between the highly fragmented, theoretical, and often not readily available information in the scientific literature and the product-specific information available from suppliers. It enlisted experts in specific areas to contribute their expertise to the development and fruition of this series.

The content of the books has been prepared in a rigorous manner, including substantial peer review and editing, and is presented in a user friendly format with definitions of terms, examples, illustrations, and trouble-shooting tips. The result is a set of practical guides containing information useful to those involved in product development, production, testing, ingredient purchasing, engineering, and marketing aspects of the food industry.

Acknowledgment of Sponsors for *Starches*

Eagan Press thanks the following companies for their financial support of this handbook:

Cerestar USA, Inc.
Hammond, IN
800/348-9896

Manildra Milling Corp.
Shawnee Mission, KS
913/362-0777

National Starch and Chemical Co.
Bridgewater, NJ
800/797-4992

A. E. Staley Manufacturing Company
Decatur, IL
800/526-5728

Eagan Press has designed this handbook series as practical guides serving the interests of the food industry as a whole rather than the individual interests of any single company. Nonetheless, corporate sponsorship has allowed these books to be more affordable for a wide audience.

Acknowledgments

Eagan Press thanks the following individuals for their contributions to the preparation of this book:

Chung-Wai Chiu, National Starch and Chemical Company, Bridgewater, NJ

Mark Hanover, A. E. Staley Manufacturing Company, Decatur, IL

David Huang, National Starch and Chemical Company, Bridgewater, NJ

Thomas Luallen, ConAgra, Omaha, NE

Amy Nelson

Patricia A. Richmond, A. E. Staley Manufacturing Company, Decatur, IL

Eric R. Shinsato, Cerestar USA, Inc.

Pepipa Yue, National Starch and Chemical Company, Bridgewater, NJ

Contents

CHAPTER 1

Starch Structure

In This Chapter:

Starch is the primary source of stored energy in cereal grains. Although the amount of starch contained in grains varies, it is generally between 60 and 75% of the weight of the grain and provides 70–80% of the calories consumed by humans worldwide.

In addition to their nutritive value, starches and modified starches can be used to affect the physical properties of many foods. For example, commercial starches obtained from corn, wheat, various rices, and tubers such as potato, sweet potato, and cassava (tapioca starch) can be used in gelling, thickening, adhesion, moisture-retention, stabilizing, texturizing, and antistaling applications. Starch and products derived from starch are also important in the paper and textile industries. The unique chemical and physical characteristics of starch set it apart from all other carbohydrates.

Basic Carbohydrate Chemistry

Regardless of the botanical source, starch is basically polymers of the six-carbon sugar D-glucose, often referred to as the "building block" of starch. The structure of the monosaccharide D-glucose can be depicted in either an open-chain or a ring form (Fig. 1-1). The ring configuration is referred to as a pyranose, i.e., *D-glucopyranose*. The pyranose ring is the most thermodynamically stable and is the configuration of the sugar in solution. The highly reactive aldehyde group at carbon number 1 (C1) on D-glucose makes it a *reducing sugar*. In biological systems, D-glucopyranose is usually present in relatively small amounts compared with the levels of various disaccharides and polysaccharides that are present.

Starch consists primarily of D-glucopyranose polymers linked together by α-1,4 and α-1,6 *glycosidic bonds* (Fig. 1-2). In forming these bonds, carbon number 1 (C1) on a D-glucopyranose molecule reacts with carbon number 4 (C4) or carbon number 6 (C6) from an adjacent D-glucopyranose molecule. Because the aldehyde group on one end of a starch polymer is always free, starch polymers always have one reducing end. The other end of the polymer is called the nonreducing end. Depending on the number of polymeric branches present in a starch molecule, there could be a large number of nonreducing ends. The glycosidic linkages in starch are in the alpha (α) configuration. Formation of an α linkage is determined by the orien-

D-Glucopyranose—The ring form of the monosaccharide D-glucose.

Reducing sugar—A monosaccharide, disaccharide, oligosaccharide, or related product capable of reducing an oxidizing ion. A common test for the measurement of reducing sugars involves the reduction of cupric ions (Cu^{+2}) to cuprous ions (Cu^{+}).

Glycosidic bond—Covalent linkage formed between D-glucopyranose units.

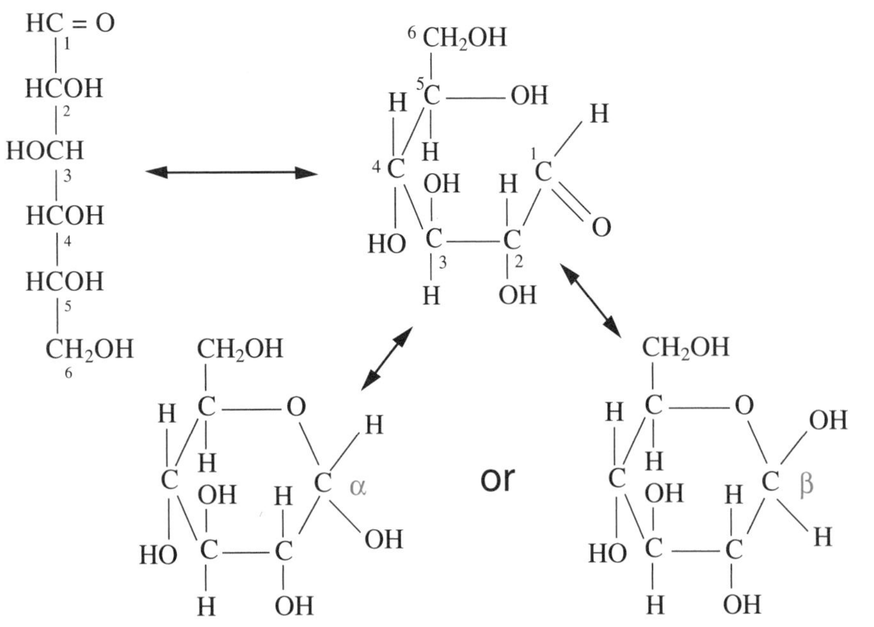

Fig. 1-1. Open-chain and pyranose ring structures of the hexose sugar D-glucose. The ring form is referred to as D-glucopyranose and can be in either the α or β configuration. Starch polymers contain only α linkages.

tation of the hydroxyl (–OH) group on C1 of the pyranose ring (Fig. 1-1). The α linkage allows some starch polymers to form helical structures. The importance of the helical geometry of starch polymers is discussed at various times throughout this text. To illustrate the significance of the α linkage, starch is sometimes compared to *cellulose,* a glucose polymer with β-1,4 bonds between subunits. This seemingly trivial difference results in large differences between starch and cellulose polymers, most notably in structural configuration, physicochemical properties, and susceptibility to certain enzymes. Because of its β configuration, cellulose forms a sheeted, ribbon-like structure, whereas starch polymers are usually helical. The α configuration and the helical geometry of starch contribute to its unique properties and enzyme digestibility. Starch polymers can be hydrolyzed by *amylase* enzymes, often referred to as the "starch-splitting" enzymes (see Chapter 4). Since the β-1,4 bonds of cellulose are not susceptible to amylase enzymes, cellulose cannot be digested by most animals.

Glucose polymerization in starch results in two types of polymers, *amylose* and *amylopectin*. Amylose is an essentially linear polymer, whereas the amylopectin molecule is much larger and is branched. The structural differences between these two polymers contribute to significant differences in starch properties and functionality.

Cellulose—The most abundant carbohydrate polymer on earth, found only in plants, comprising β-1,4-linked D-glucopyranose units.

Amylase—Any one of several starch-degrading enzymes common to animals, plants, and microorganisms.

Amylose—An essentially linear polymer of starch composed of α-1,4-linked D-glucopyranose molecules. A small number of α-1,6-linked branches may be present.

Amylopectin—A very large, branched, D-glucopyranose polymer of starch containing both α-1,4 and α-1,6 linkages. The α-1,6 linkage represents the bond at the polymeric branch point.

Starch Polymer Biosynthesis

Starch functions mainly as a carbohydrate source for the growing plant (e.g., for germinating seeds and leaf tissue development) and is consequently the primary source of stored energy in the plant. Depending upon the plant, starch can be found in a variety of tissues, including leaves, tubers, fruits, and seeds. Although a significant amount of information exists in the scientific literature on starch biosynthesis in higher plants, it is not completely understood and continues to represent a major area of ongoing research. A detailed description of starch biosynthesis in the plant is a complex process and is beyond the scope of this book.

In general terms, starch polymers are produced within the plastids of the plant cell by a series of complex biosynthetic pathways controlled by key enzymes. Starch synthesis is localized in the chloroplasts of green photosynthetic tissue and/or in the *amyloplasts* of non-green storage tissues. Enzymes catalyze the addition of D-glucopyranose molecules onto a growing D-glucopyranose chain, often referred to as a glucan chain. This growth includes elongation of amylose as well as formation of branches on the amylopectin molecule.

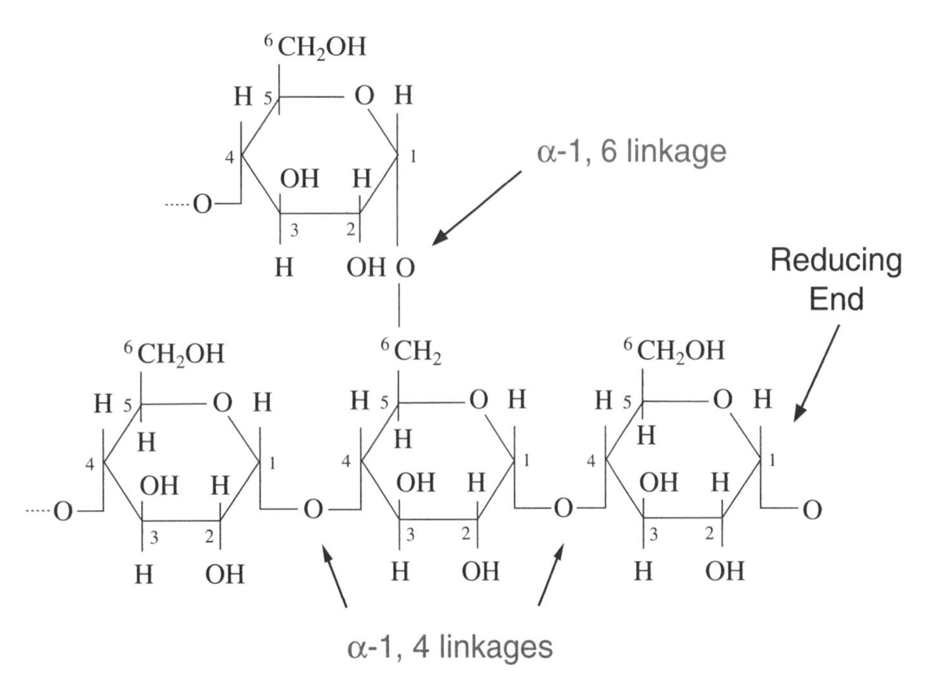

Fig. 1-2. α-1,4 and α-1,6 glycosidic bonds of starch.

Starch synthase is the enzyme that catalyzes the addition of adenosine-diphosphoglucose (ADP-glucose), a reactive form of D-glucopyranose in the plant cell, onto a growing amylose chain. Starch synthase elongates the amylose chain by successive addition of D-glucopyranose. The branched polymer amylopectin is formed as branching enzymes catalyze the addition of α-1,4 glucan chains onto existing α-1,4 glucans via α-1,6 linkages at the branch points. It is believed that these branching enzymes utilize a "cut-and-paste" mechanism that positions new "branches" onto existing α-1,4 glucans (1).

The various biosynthetic reactions responsible for starch polymer synthesis are under a great deal of metabolic control, and the exact pathway that results in polymer formation is still not completely understood. Given the complexity of starch biosynthesis, it is easy to see why amylose and amylopectin polymers vary in size and structure depending on the plant and its metabolic requirements. These inherent differences between starches from various sources contribute to their versatility as food ingredients.

Properties of Amylose and Amylopectin

Amyloplasts—Organelles in plant cells that synthesize starch polymers in the form of granules.

Although amylose and amylopectin are both composed of D-glucopyranose molecules, dissimilarities between these two poly-

TABLE 1-1. Characteristics of Amylose and Amylopectin

Characteristic	Amylose	Amylopectin
Shape	Essentially linear	Branched
Linkage	α-1,4 (some α-1,6)	α-1,4 and α-1,6
Molecular weight	Typically <0.5 million	50–500 million
Films	Strong	Weak
Gel formation	Firm	Non-gelling to soft
Color with iodine	Blue	Reddish brown

mers result in major differences in functional properties. Some important characteristics of amylose and amylopectin are listed in Table 1-1.

Hydrophobic—"Water-fearing." The term is usually used to describe nonpolar substances, e.g., fats and oils, that have little or no affinity for water.

Clathrate—An inclusion complex wherein a "host" molecule entraps a second molecular species as the "guest."

Mono- and **Diglycerides**—Glycerol molecules with one or two fatty acids attached, respectively.

AMYLOSE

Amylose is considered to be an essentially linear polymer composed almost entirely of α-1,4-linked D-glucopyranose (Fig. 1-3). Recent evidence, however, has suggested that some branches are present on the amylose polymer (2). Simplified models for the structure of amylose are shown in Figure 1-4. Although typically illustrated as a straight chain structure for the sake of simplicity, amylose is actually often helical. The interior of the helix contains hydrogen atoms and is therefore *hydrophobic,* allowing amylose to form a type of *clathrate* complex with free fatty acids, fatty acid components of glycerides, some alcohols, and iodine (3).

Iodine complexation is an important diagnostic tool for the characterization of starch and is reviewed in greater detail in Chapter 2. Complexation with lipids, particularly *mono-* and *diglycerides,* is a well-known property of the amylose helix. The formation and structural integrity of amylose-lipid complexes are functions of various factors, including temperature, pH, contact and/or mixing time between the host amylose polymer and the "guest" molecule, and the structure of the fatty acid or glyceride. The resulting "inclusion com-

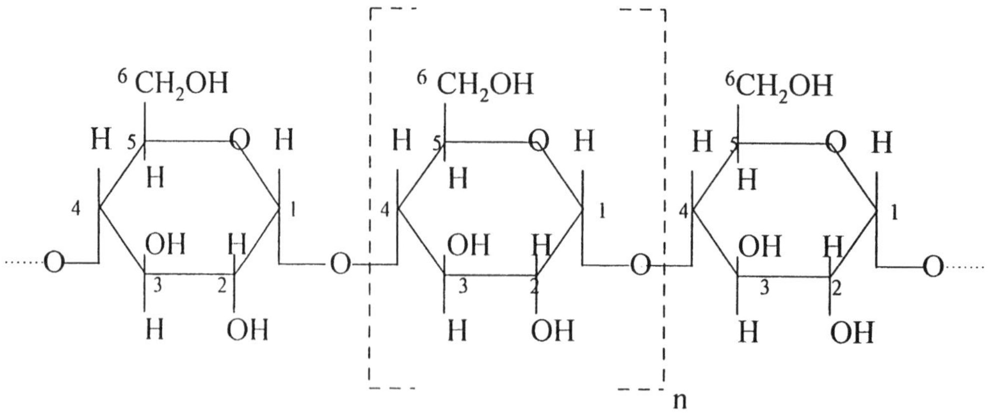

Fig. 1-3. α-1,4 linkages of amylose.

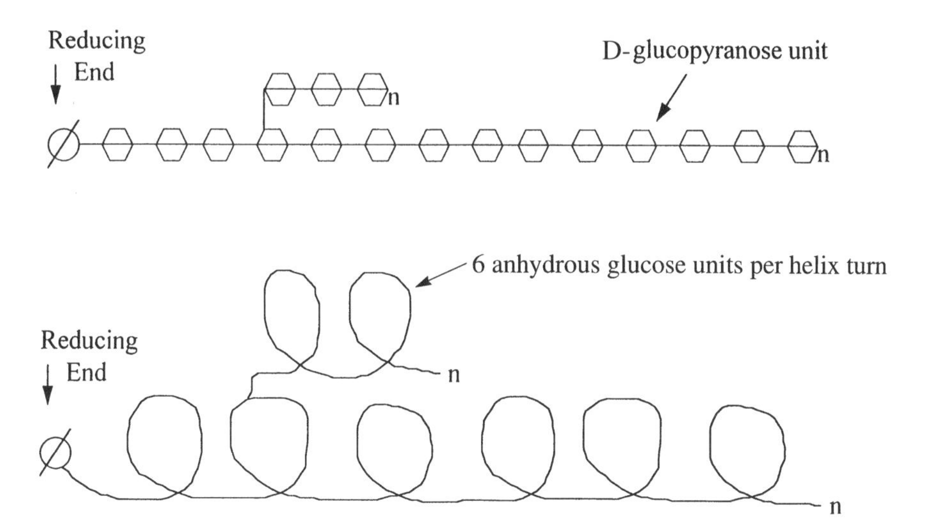

Fig. 1-4. Amylose models. Amylose can be depicted as either a straight chain or a helix.

Gelatinization—Collapse (disruption) of molecular orders within the starch granule manifested by irreversible changes in properties such as granular swelling, native crystalline melting, loss of birefringence, and starch solubilization.

Paste—Starch in which a majority of the granules have undergone gelatinization, giving it a viscosity-forming ability. Pasting involves granular swelling and exudation of the granular molecular components.

Retrogradation—Process during which starch chains begin to reassociate in an ordered structure. Two or more starch chains initially form a simple juncture point, which then may develop into more extensively ordered regions and ultimately, under favorable conditions, to a crystalline order.

Corn starch—Common corn starch composed of approximately 25% amylose and 75% amylopectin.

High-amylose corn starch—Starch isolated from a hybrid corn plant that contains greater than about 40% amylose. Some high-amylose corn starches now contain as much as 90% amylose.

plex" (Fig. 1-5), as it is often called, can alter the properties of the starch. As depicted, the hydrophobic core of the amylose helix complexes with the hydrophobic constituent. Amylose complexation with fats and food emulsifiers such as mono- and diglycerides can shift starch *gelatinization* temperatures, alter textural and viscosity profiles of the resultant *paste*, and limit *retrogradation*. (Gelatinization, pasting, and retrogradation are discussed in Chapter 3.)

Another well-known attribute of amylose is its ability to form a gel after the starch granule has been cooked, i.e., gelatinized and pasted. This property is evident in the behavior of certain amylose-containing starches. *Corn starch*, wheat starch, rice starch, and particularly *high-amylose corn starch* isolated from hybrid corn plants are usually considered gelling starches. Gel formation is primarily the result of the reassociation (i.e., retrogradation) of solubilized starch polymers after cooking and can occur quite rapidly with the linear polymer amylose.

AMYLOPECTIN

The literature proposes several models for helical configurations, branch chains, cluster patterns, and molecular dimensions of amylopectin. The evolution of the amylopectin model has progressed with the increasing sophistication of biochemical techniques.

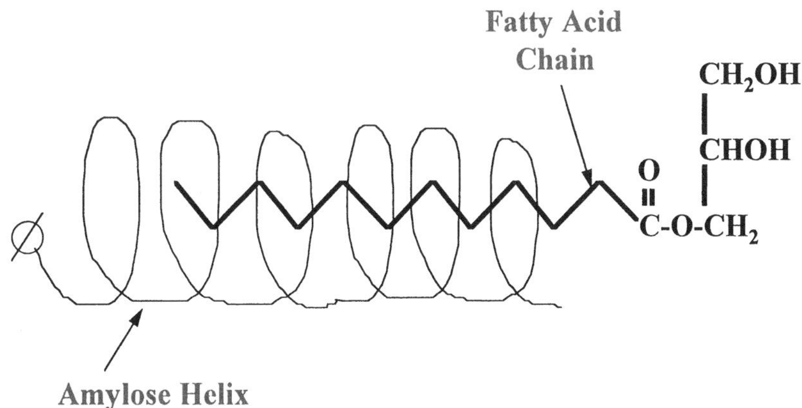

Fig. 1-5. Starch-lipid inclusion complex. An amylose helix is complexed with the fatty acid chain of a monoglyceride.

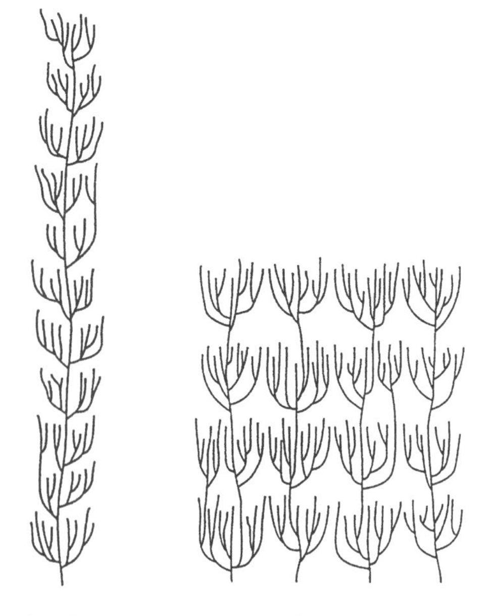

Fig. 1-6. Representation of a portion of an amylopectin molecule (left) and enlargement of typical packed clusters (right). Individual chains are helical, and pairs of chains are double helical. (Reprinted from Whistler, R. L., and BeMiller, J. N., 1997, *Carbohydrate Chemistry for Food Scientists,* American Association of Cereal Chemists, St. Paul, MN)

Amylopectin, the predominant molecule in most normal starches, is a branched polymer that is much larger than amylose (Fig. 1-6). Amylopectin is composed of α-1,4-linked glucose segments connected by α-1,6-linked branch points. It has been estimated that about 4–6% of the linkages within an average amylopectin molecule are α-1,6 linkages (4). This may appear to be a small percentage, but it results in more than 20,000 branches in an average molecule, although the branches themselves are not large. Studies suggest a bimodal size distribution of polymer chains, namely small and large chains (5,6). The small chains have an average *degree of polymerization (DP)* of about 15, whereas that of the larger chains is about 45. This unique configuration contributes to the crystalline nature of amylopectin and an ordered arrangement of amylopectin molecules within the starch granule. The behavior of branch chains of amylopectin is similar to that of amylose chains in that entire chains or, more commonly, portions of chains can be helical.

Because of the highly branched nature of amylopectin, its properties differ from those of amylose. For example, given the size of the molecule and its "tumbleweed-like" structure, retrogradation is slowed and gel formation can either be delayed or prevented. Pastes from starches that contain essentially all amylopectin (*waxy starches*) are considered to be non-gelling but typically have a cohesive and gummy texture.

Amylose from various botanical sources has a DP of about 1,500–6,000. The much larger amylopectin molecule has a DP of about 300,000–3,000,000 (7). On the basis of these numbers and given a molecular weight (MW) for anhydroglucose of 162, the MW of amylose can range from about 243,000 to 972,000. Although amylose from potato starch has been reported to have a MW of up to about 1,000,000, the MW of amylose is typically less than 500,000. The MW of amylopectin can range from about 10,000,000 to

Degree of polymerization (DP)—The molecular size of a polymer. In this case, it refers to the number of α-1,4-linked D-glucopyranose units in a starch chain.

Waxy starch—"Amylopectin" starch from certain cereals such as corn and rice. The amylopectin content of waxy starches is generally about 95% or more. The term "waxy" has nothing to do with the presence of wax, but rather describes the appearance of the cross section of the mutant corn kernel from which it was first isolated.

TABLE 1-2. Approximate Amylose and Amylopectin Content of Common Food Starches

Starch Type	Amylose Content (%)	Amylopectin Content (%)
Dent corn	25	75
Waxy corn	<1	>99
Tapioca	17	83
Potato	20	80
High-amylose corn	55–70 (or higher)	45–30 (or lower)
Wheat	25	75
Rice	19	81

500,000,000. Differences in the MW of amylose and amylopectin fractions are directly related to the plant source, the method of polymer isolation (usually a solvent precipitation method is used), and the method of MW determination.

The ratio of amylose to amylopectin within a given type of starch is a very important point to consider with respect to starch functionality in foods. The amylose and amylopectin content and structure affect the architecture of the starch granule, gelatinization and pasting profiles, and textural attributes. The approximate amylose and amylopectin contents of several starches are shown in Table 1-2. By using classical breeding techniques as well as sophisticated molecular biology, it is now possible to obtain starches from various hybrid plant sources that contain essentially all amylose, all amylopectin, or various ratios in between.

Starch granules—Naturally occurring, partially crystalline, discrete aggregates of amylose and amylopectin.

Compound starch granules—Naturally occurring aggregates of individual starch granules.

Starch Granules

Amylose and amylopectin do not exist free in nature, but as components of discrete, semicrystalline aggregates called *starch granules*. The size, shape, and structure of these granules vary substantially among botanical sources (Table 1-3). The diameters of the granules generally range from less than 1 µm to more than 100 µm, and shapes can be regular (e.g., spherical, ovoid, or angular) or quite irregular. Wheat, barley, and rye granules exhibit two separate distributions of granule sizes and distinctly different shapes. For example, large, oval granules (type A) of approximately 35 µm (major axis) and smaller, more spherical granules (type B) about 3 µm in diameter can both be extracted from wheat flour. Some granules, such as those in oats and rice, have a higher level of structure in which many small, individual granules are cohesively bound together in an organized manner. These are called *compound starch granules*. Although the major components of all types of starch granules are amylose and amylopectin polymers, there is great diversity in the structure and characteristics of native starch granules.

TABLE 1-3. Approximate Size and Shape of Common Food Starch Granules[a]

Property	Dent Corn	Waxy Corn	High-Amylose Corn	Wheat	Rice	Potato	Tapioca
Source	Cereal	Cereal	Cereal	Cereal	Cereal	Tuber	Root
Diameter (µm)	5–30	5–30	5–30	1–45	1–3	5–100	4–35
Shape	Polygonal, round	Polygonal, round	Polygonal, round, irregular	Round, lenticular	Polygonal, spherical compound granules	Oval, spherical	Oval, truncated, "kettle drum"

[a] Adapted from Alexander, R. J., 1995, Potato starch: New prospects for an old product, Cereal Foods World 40:763–764.

INTERNAL STRUCTURE OF THE STARCH GRANULE

The arrangement of amylose and amylopectin within the starch granule is not completely understood. The "packaging" of these two polymers in the native starch granule is not random but is very organized. When heated in the presence of water, however, starch granules become much less ordered. This loss of internal order occurs at different temperatures for different types of starch. Depending on the starch, if it is heated in water indefinitely, the native granule swells until its structure eventually disintegrates, and amylose and amylopectin are released into an aqueous suspension.

Much of what is known about the internal structure is the result of microscopic evaluation of partially degraded granules. Compared with crystalline areas, amorphous areas of the granule are generally degraded more easily by acid and enzymes, such as α-amylase. Granules exhaustively treated in this manner demonstrate a ringed pattern analogous to the growth rings on a crosscut piece of wood from a tree trunk (Fig. 1-7). This pattern indicates that the crystalline and amorphous areas of the granule alternate. It is thought that this configuration results from alternating periods of growth and rest during the synthesis of the starch granule (8).

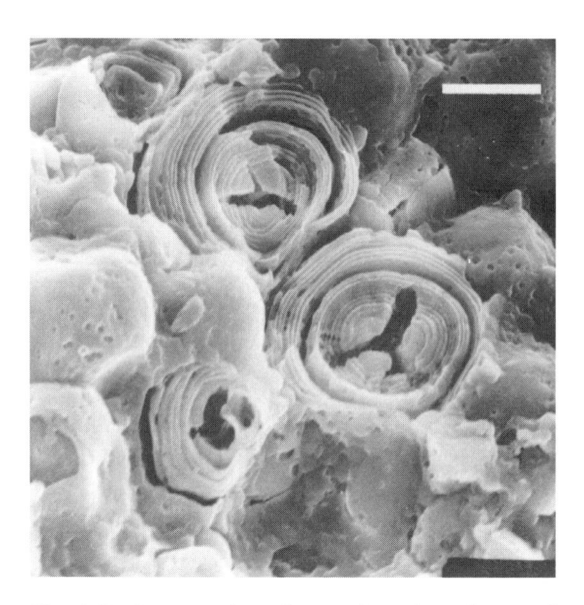

Fig. 1-7. Cross section of a sorghum kernel treated with α-amylase. Note the concentric rings in the broken starch granules. Bar = 10 μm. (Reprinted from [8])

As discussed previously, amylopectin is a large molecule composed of two distinct populations of chain lengths. The smaller chains are thought to be in such close proximity that they interact strongly, resulting in crystalline regions that are quite extensive and arranged regularly with respect to each other throughout the granule. The model in Figure 1-8, describing the arrangement of the amylopectin molecule within a growth ring of a starch granule, has been proposed by French and Kainuma (9).

These radially oriented amylopectin "clusters" are also believed to be associated with amylose, which is interwoven throughout the crystalline and amorphous areas. The location of amylose within the granule remains one of the unknown facts required to complete our picture of the internal structure of the starch granule.

MINOR CONSTITUENTS OF THE STARCH GRANULE

Proteins, lipids, moisture, and ash (minerals and salts) are also present in starch granules in very small quantities. Moisture typically equilibrates to about 12% in a starch powder. Ash content, although variable, is typically less than 0.5% (dry basis). Table 1-4 summarizes the lipid and protein contents of common food starches. The values reported are not limited to the proteins and lipids associated with the granules but reflect residuals from the starch isolation (milling) process.

Generally speaking, tuber and root starches contain less lipid and protein compared with cereal starches. Although minor amounts of residual lipid and protein can influence gelatinization, the most dramatic effect of these components is on the flavor profile of the starch. Compared with most cereal starches, tapioca and potato starch are considered to be very bland in flavor because of the small amounts of lipid and protein present. Although some starch sources are inherently "cleaner" with respect to their lipid and protein contents, in most instances the levels of these components are directly related to the process used to isolate the granules. Because starch milling is generally a wet fractionation and purification process, the amount of residual lipid and protein is a function of the efficacy of the milling (starch isolation) process. Novel milling techniques, such as an alkali washing step in the milling of corn starch, can significantly reduce the levels of residual lipid and protein.

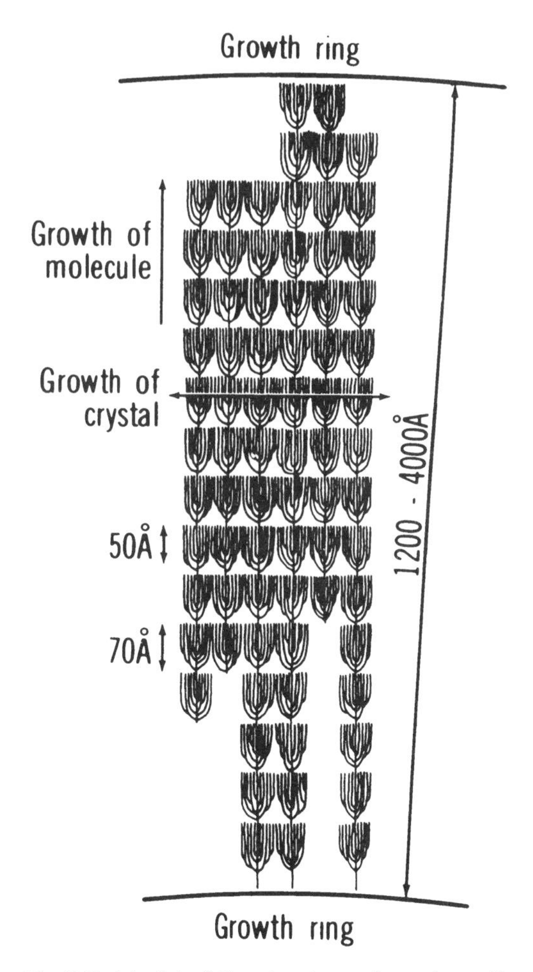

Fig.1-8. Model of the structure of amylopectin in starch granules (1 Å = 0.1 nm). (Reprinted, with permission, from [9])

Proteins. Starch granule proteins have been divided into two types on the basis of their ability to be extracted from the granules (10). "Surface" starch granule proteins can be extracted with salt solutions, whereas "integral" starch granule proteins require more rigorous extraction, for example, with the detergent sodium dodecyl sulfate or an alkaline solution. It has been theorized that the integral proteins are embedded and possibly covalently bound in the amylose-amylopectin structure of the granule, while the surface proteins are more loosely associated with the exterior of the granule.

The presence of a specific starch granule protein has been demonstrated to correlate well with the *hardness* of wheat *endosperm*. Greenwell and Schofield, after extraction of starch granule proteins in sodium dodecyl sulfate, isolated a specific protein they called the 15K starch granule protein (11). This protein was much more abundant in starch extracted from soft wheat than in that from hard wheat. This study involved a genetically diverse set of more than 150 wheat cultivars and strongly suggests that starch granule proteins play a significant role in determining hardness.

Hardness—Amount of force required to crush a kernel of wheat. Hard wheats require more energy to mill to flour and generate more damaged starch during the process than soft wheats.

Endosperm—Interior portion of the wheat kernel containing the gluten and starch, which comprise flour; the fuel source for a sprouting wheat plant.

TABLE 1-4. Approximate Lipid and Protein Content of Common Food Starches[a]

	Dent Corn	Waxy Corn	Wheat	Potato	Tapioca
Protein (%)	0.35	0.25	0.4	0.1	0.1
Lipid (%)	0.80	0.20	0.90	0.1	0.1

[a] Adapted from Davies, L., 1995, Starch—Composition, modifications, applications and nutritional value in foodstuffs, Food Tech Europe, June/July, pp. 44-52.

Germ—Embryo of the wheat plant within the kernel.

Spherosomes—Fat-containing organelles in endosperm cells.

Triglyceride—Three fatty acids attached to a glycerol molecule.

Ash—Mineral and salt fraction, typically calculated by determining the amount of residue remaining after incineration of a sample.

Lipids. Most of the lipid components of cereal grains are concentrated in the *germ*. The lipids of the endosperm have been classified as starch lipids (those associated with starch granules) and non-starch lipids (such as those contained in the *spherosomes* dispersed throughout the endosperm). The non-starch lipids have diverse structures; *triglycerides* and other non-polar lipids are the major components.

Starch lipids are contained within the granules. These lipids contain phosphate groups and are similar in structure to the common emulsifier lecithin, with one fatty acid group per molecule. They are termed "lysophospholipids," and their structure allows them to form a complex with amylose wherein the fatty acid group is aligned in the core of the amylose helix (Fig. 1-5). This complex is very stable and dissociates or "melts" only at very high temperatures.

How the lysophospholipids are arranged in the starch granule is a matter of some conjecture. It is generally thought that they are highly associated with the amylose components of the starch granules (12). There is some evidence, however, that much of the lysophospholipid content is not contained within amylose inclusion complexes. For example, Biliaderis et al. (13) generated thermal data indicating that many of the amylose complexes do not form until the starch is gelatinized.

Ash. Starches also contain trace amounts of mineral elements and inorganic salts. For example, tuber starches contain covalently bound phosphate. Collectively, these minerals and salts are referred to as *ash*. Ash content can vary depending upon the source of the raw material, agronomic practices, milling procedures, and types of chemical modifications that may be done to the starch. The ash content of starch is typically less than 0.5% of the dry matter.

References

1. Wasserman, B. P., Harn, C., Mu-Forster, C., and Huang, R. 1995. Biotechnology: Progress toward genetically modified starches. Cereal Foods World 40:810-817.
2. Curá, J. A., Jansson, P.-E., and Krisman, C. R. 1995. Amylose is not strictly linear. Starch/Staerke 47:207-209.
3. Fennema, O. R. 1985. Water and ice. Pages 23-67 in: *Food Chemistry*. O. R. Fennema, Ed. Marcel Dekker, New York.
4. Hood, L. F. 1982. Current concepts of starch structure. Pages 218-224 in: *Food Carbohydrates*. D. R. Lineback and G. E. Inglett, Eds. AVI, Westport, CT.
5. Robin, J. P., Mercier, C., Charbonniere, R., and Guilbot, A. 1974. Lintnerized starches: Gel filtration and enzymatic studies of insoluble residues from prolonged acid treatment of potato starch. Cereal Chem. 51:389-406.
6. Hizukuri, S. 1986. Polymodal distribution of the chain lengths of amylopectins, and its significance. Carbohydr. Res. 147:342-347.
7. Zobel, H. F. 1988. Molecules to granules: A comprehensive starch review. Starch/Staerke 40:44-50.
8. Hoseney, R. C. 1994. *Principles of Cereal Science and Technology*, 2nd ed. American Association of Cereal Chemists, St. Paul, MN.

9. French, D. 1984. Organization of starch granules. Pages 183-247 in *Starch: Chemistry and Technology*, 2nd ed. R. L. Whistler, J. N. BeMiller, and E. F. Paschall, Eds. Academic Press, New York.
10. Rayas-Duarte, P., Robinson, S. F., and Freeman, T. P. 1995. In situ location of a starch granule protein in durum wheat endosperm by immunocytochemistry. Cereal Chem. 72:269-274.
11. Greenwell, P., and Schofield, J. D. 1986. A starch granule protein associated with endosperm softness in wheat. Cereal Chem. 63:379-380.
12. Acker, L. 1982. The role of starch lipids among cereal lipids, their composition and their importance for the baking properties of wheat flours. Getreide Mehl Brot 36:291-295.
13. Biliaderis, C. J., Page, C. M., and Maurice, T. J. 1986. On the multiple melting transitions of starch/monoglyceride systems. Food Chem. 22:279-295.

CHAPTER

2

Starch Analysis Methods

In This Chapter:

Starch appears as a white, powdery material to the naked eye. Over the years, however, instruments and methods have been developed that have led to a better understanding of the basic structure of starch and of the changes that take place in the presence of water, heat, or other food ingredients.

Microscopic analysis of starch has been used to study the structure of the granule and X-ray crystallography its crystalline structure. Instruments are available that measure the changes in viscosity that take place as starch solutions are heated; and procedures have been developed for determining swelling power, the starch content of cereal grains and food products, the level of starch damaged during the extraction and refining processes, and the amylose and amylopectin contents of starch samples. Techniques have also been developed for quantitating and evaluating gelatinization.

Microscopy

Light and electron microscopy are valuable tools in the characterization of starch and starch-containing food products. Much of what is known about granular structure was determined by microscopic means.

LIGHT MICROSCOPY

Light microscopy is often used to identify the type of starch. The first person to observe starch granules was the inventor of the light microscope, Antonie van Leeuwenhoek, in 1719. The general size and shape of starch granules from different sources can be observed with this technique. Light micrographs of wheat, potato, corn, and rice starches appear in Figure 2-1. Wheat starch is shown in Figure 2-1a, and it can clearly be seen that there are two populations of granule sizes. The large granules are about 20–45 μm in diameter (A granules), and the smaller granules are about 1–5 μm in diameter (B granules). The larger, oval potato starch granules in Figure 2-1b can be more than 100 μm along their major axis. Corn starches, such as that shown in Figure 2-1c, have round to polygonal granules about 5–30 μm in diameter. All of these starches exist as individual granules. In contrast, Figure 2-1d shows the compound granules of rice starch made up of hundreds of small, individual granules, each one

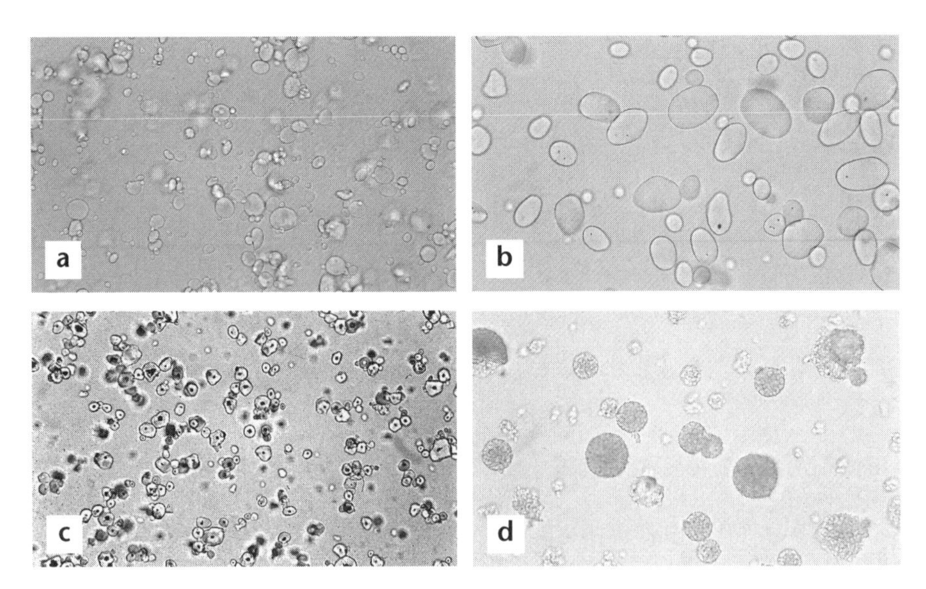

Fig. 2-1. Light micrographs of starches from a, wheat; b, potato; c, corn; and d, rice.

about 1–3 μm in diameter. Tapioca starch (not shown) has a truncated (acornlike) shape, often referred to as a "kettle drum."

Polarized light microscopy. When viewed with a microscope under polarized light, all native starch granules appear to shine while exhibiting a dark "Maltese cross." This phenomenon is known as *birefringence* and is an indicator of the degree of order in the granules. As native starch granules are heated in the presence of water, they lose their birefringence. The loss of birefringence is closely associated with the phenomenon known as "gelatinization" (see Chapter 3) and, in fact, is often used as an indicator of gelatinization.

The Maltese cross is clearly apparent in both the A and B granules of wheat starch (Fig. 2-2a), potato starch (Fig. 2-2b), and corn starch (Fig. 2-2c). The compound granules of rice (Fig. 2-2d) are large and shiny when viewed under polarized light. However, careful examination of Figure 2-2d reveals that the individual rice starch granules also possess the birefringent cross.

Although the procedure is somewhat tedious, it is possible to use a light microscope equipped with a polarizer and a heating stage to quantitate the loss of birefringence of a starch sample. Starch is first dispersed in water and fixed on a microscope slide. By slowly heating this slide while counting the birefringent granules in a specific field of starch granules, data relating temperature to a percent birefringence can be obtained. Generally, the loss of birefringence occurs over a temperature range.

Birefringence—Phenomenon that occurs when polarized light interacts with a highly ordered structure, such as a crystal. A crossed diffraction pattern, often referred to as a "Maltese cross," is created by the rotation of polarized light by a crystalline or highly ordered region, such as that found in starch granules.

Light microscopy with iodine staining. Another well-known technique for the microscopic evaluation of starch employs iodine staining to provide information about the amylose and amylopectin content. Depending on the amylose content, starches, particularly intact

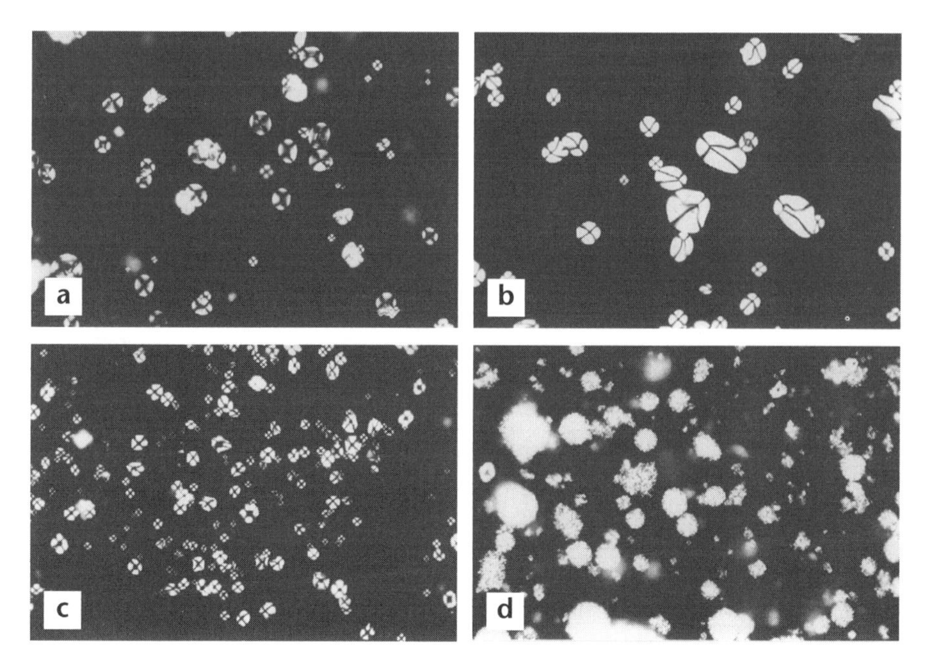

Fig. 2-2. Starches from a, wheat; b, potato; c, corn; and d, rice under polarized light.

A usable recipe for an iodine staining solution consists of mixing about 0.6 g of potassium iodide and 0.4 g of iodine into 2 ml of water. The solution is then diluted to 500 ml with water and stored in a covered container in a dark location (2).

granules, stain blue or reddish brown in the presence of a solution of iodine and potassium iodide. The resultant color is dependent upon the complex formed when iodine is enclosed in helical starch chains within the granular structure (1). When the helix is long, the complex is blue. Complexes with shorter chains are purple or reddish brown. Branch points in a starch polymer, such as those present in amylopectin, disrupt this helical structure. Therefore, amylopectin mixed with an iodine-potassium iodide solution stains reddish brown. The more helical structure of amylose results in a blue complex. Corn starch contains about 25% amylose and stains blue. Waxy starches, which contain essentially 100% amylopectin, stain reddish brown.

Figure 2-3 contains micrographs of three corn starches with different amylose-amylopectin ratios as well as normal and waxy rice starch. Waxy corn starch (Fig. 2-3a) does not bind much iodine and appears red under the light microscope. Both dent and high-amylose corn starch bind large amounts of iodine and stain blue (Fig. 2-3b and c, respectively). Similarly, both the individual and compound granules of normal rice appear blue (Fig. 2-3d). The waxy rice shown in Figure 2-3e appears reddish brown.

SCANNING ELECTRON MICROSCOPY

Scanning electron microscopy (SEM) allows the shape and surface features of starch granules to be viewed in three dimensions. When a scanning electron microscope is used, the sample is coated with a thin layer of a reflective metal and then irradiated with a beam of

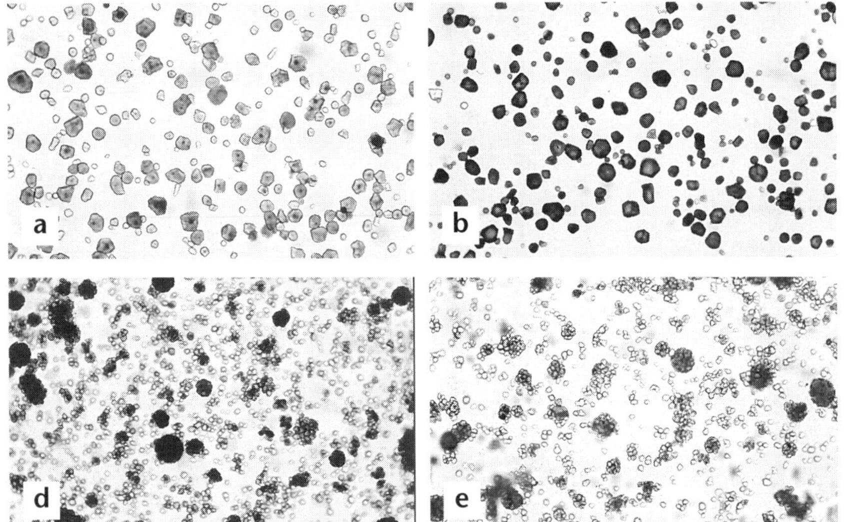

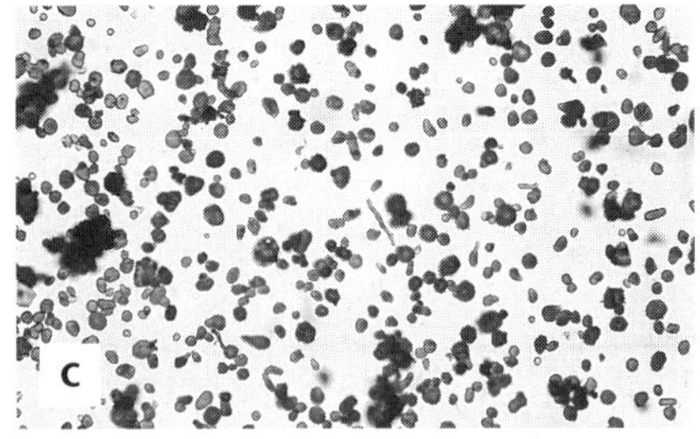

Fig. 2-3. Iodine-stained starches from a, waxy corn; b, dent corn; c, high-amylose corn; d, rice; and e, waxy rice.

electrons. The electrons reflect back to a sensor, allowing surface features of starch granules to be seen in great detail. An excellent review of starch granule SEM has been provided by Jane et al. (3).

When starch granules are viewed by SEM, a number of interesting features become apparent. The field of wheat starch granules in Figure 2-4a contains both A (large) and B (small) granules. Indentations are clearly visible on some of the A granules, showing where B granules, protein bodies, and other cellular components were packed tightly together in the wheat endosperm. One feature unique to native wheat starch A granules is the existence of an equatorial groove, which can be seen in Figure 2-4b. Dent corn starch granules (Fig. 2-4c) are angular and less rounded than those of wheat starch. Surface packing features are also visible on some of these granules. High-amylose corn starch (Fig. 2-4d) contains a population of granules with some very curious shapes. For example, the granule in the center of this micrograph has a large, tubular appendage protruding from an otherwise normally shaped granule. Potato starch granules (Fig. 2-4e) are round with a very smooth exterior and few surface markings. Tapioca starch granules (Fig. 2-4f) have a distinctive kettle drum shape. Finally, the compound granules of the rice starch (Fig. 2-4g) appear as loosely packed spheres.

As large wheat starch granules are heated in the presence of water, they undergo changes in size and shape. Granules initially swell and take the shape of a flattened disk, the result of swelling in the plane of the major axes. As the heating continues, the granules take on a puckered appearance, which likely results from swelling in the direction perpendicular to the major axes. This phenomenon is apparent in the starches suspended in excess water and subjected to various temperatures (Fig. 2-5).

Starch granules have been successfully extracted from various

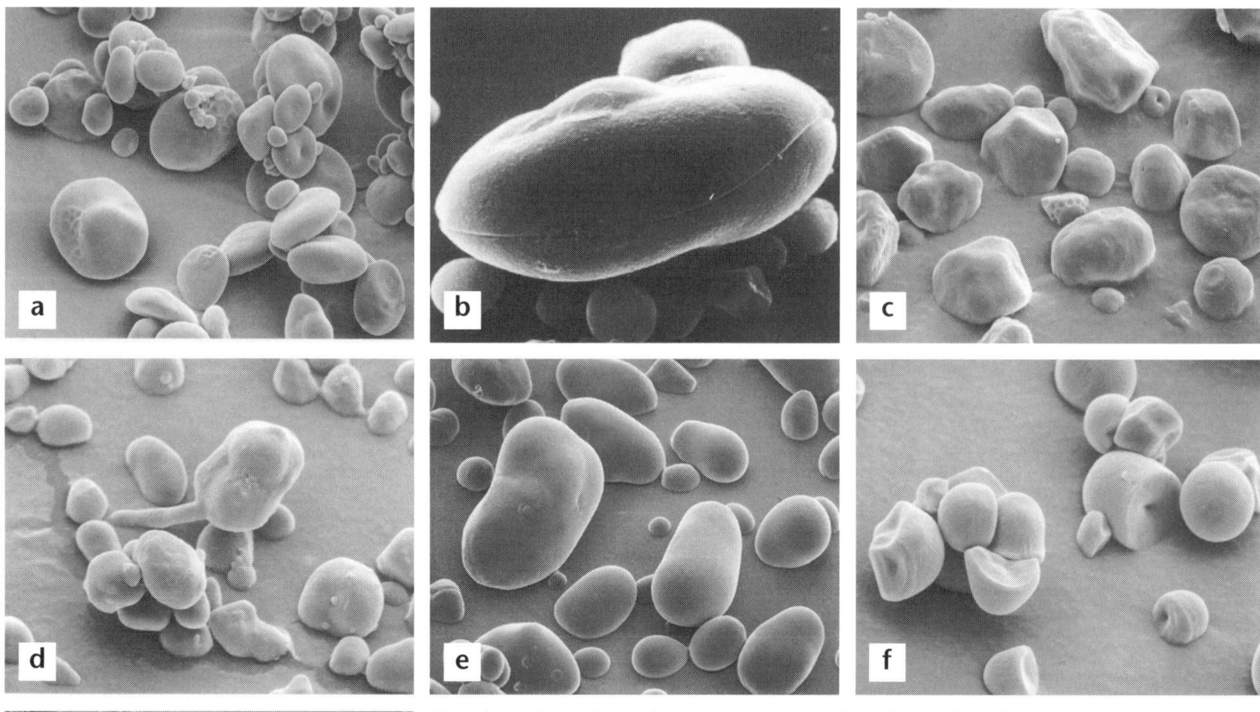

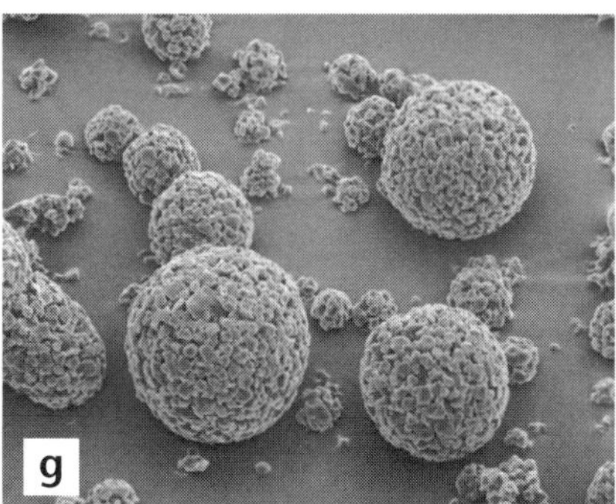

Fig. 2-4. Scanning electron micrographs of starches from a, wheat (1,000x); b, wheat; c, dent corn (2,000x); d, high-amylose corn (2,000x); e, potato (600x); f, tapioca (2,000x); and g, rice (600x). (b reprinted, with permission, from Evers, A. D., 1971, Scanning electron microscopy of wheat starch, III, Granule development in the endosperm, Starch/Staerke 23:157-162)

baked products and microscopically observed (5). Through most baking processes, wheat and corn starch granules remain intact and durable enough for extraction and SEM evaluation. Although granular integrity remains high, the degree of folding or puckering of the granules extracted from different baked-product systems does vary. This can be seen by comparing the shape of the granules extracted from pie crust (Fig. 2-6a) with those from angel food cake (Fig. 2-6b). The changes in the granules are related not only to the temperatures achieved during the baking process, but also to the water available to the starch. Granules in low-moisture systems such as pie crust do not undergo as much transformation as those in high-moisture systems such as angel food cake. Water availability for starch in these food systems is dependent not only on the amount of water added to the formula, but also on the other hygroscopic ingredients in the for-

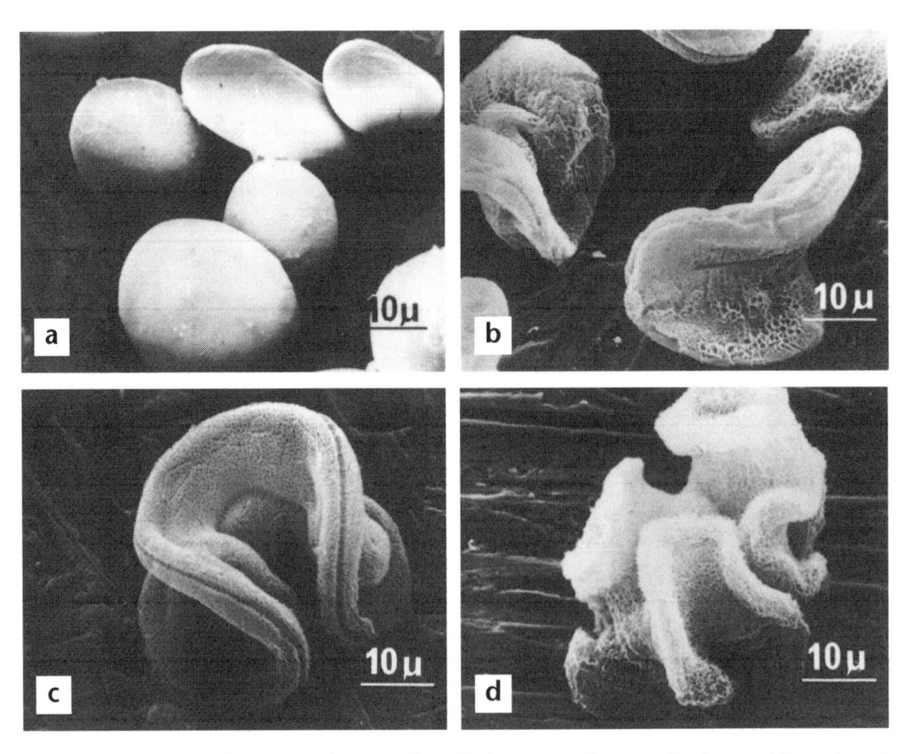

Fig. 2-5. Scanning electron micrographs of wheat starch granules heated in water at a, 20°C; b, 60°C; c, 80°C; and d, 97°C. (Reprinted, with permission, from [4])

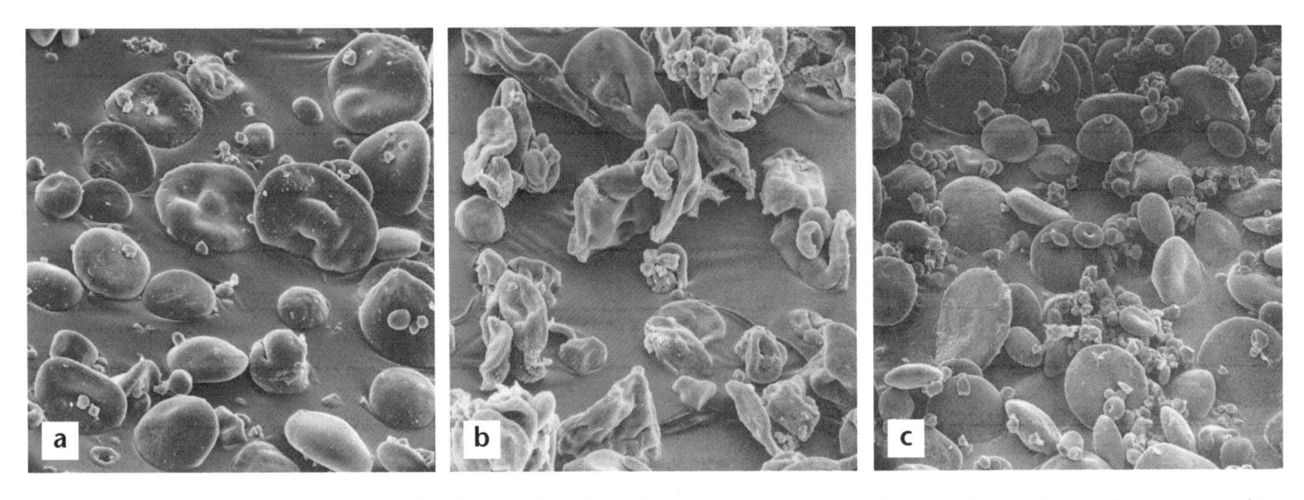

Fig. 2-6. Scanning electron micrographs of starch from baked products: a, pie crust; b, angel food cake; and c, sugar cookie. (Reprinted from [5])

mula. For example, high levels of sugar in a formula result in much less water available for starch transformation if the water percentage in the formula remains constant. In Figure 2-6c, an SEM of granules extracted from sugar cookies (an extremely high-sugar formula), the granules are essentially in their native, ungelatinized form.

X-Ray Crystallography

When a crystal is irradiated with X rays, the X rays split to form a pattern distinctive to the crystal structure. This technique has been used to study starch's crystalline nature. Three general *X-ray patterns* have been identified in native starch by this method. Native cereal starches such as wheat, corn, and rice yield an A pattern, and tuber starches such as potato yield a B pattern (Fig. 2-7). Smooth pea and bean starches give a C pattern, an intermediate form that probably results from mixtures of A and B types. If starch recrystallizes in the presence of a fatty acid or long-chain alcohol, the V pattern (Fig. 2-7) is obtained (6).

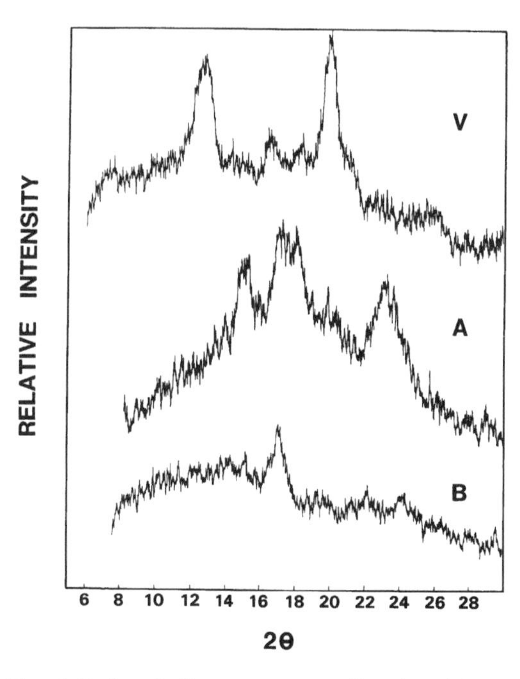

Fig. 2-7. Starch X-ray patterns. (Reprinted from [6])

On the basis of X-ray crystallography, detailed structures for starch crystals have been proposed (7). Double helices present in starch are somewhat similar to the model proposed for the structure of DNA. The crystal models for the A and B patterns vary with respect to the amount of water hydrating the glucose residues as well as to the density of the packing arrangement. The V pattern is attributed to a starch complex in which a chain of glucose residues forms a helix with a *hydrophilic* exterior and a hydrophobic core containing a nonpolar molecule such as a lipid. This type of structure is similar to the iodine-starch complex discussed previously. A general structure is shown in Figure 2-8.

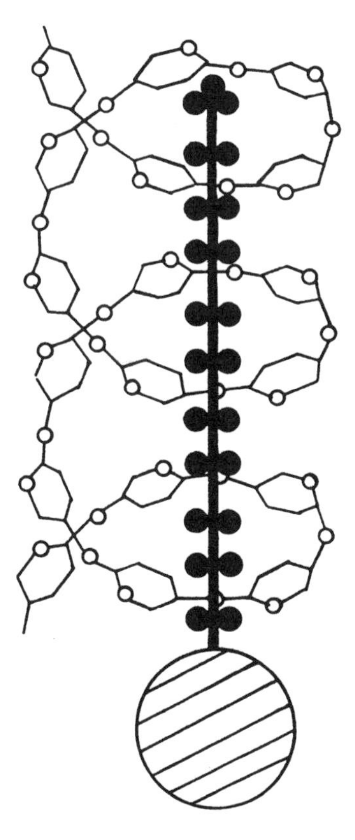

Fig. 2-8. Simplified V starch crystal structure. (Reprinted, with permission, from Carlson, T. L.-G., Larsson, K., Dinh-Nguyen, N., and Krog, N., 1979, A study of the amylose-monoglyceride complex by Raman spectroscopy, Starch/Staerke 31:222-224)

Viscosity-Based Measurements

Viscosity changes in starch media during heating are commonly measured with instruments called the viscoamylograph (C. W. Brabender, Inc., South Hackensack, NJ) and the Rapid Visco Analyser (RVA) (Newport Scientific Pty. Ltd, Warriewood, NSW, Australia). These viscometers run con-

X-ray pattern—Pattern obtained when a crystal is irradiated with X rays. This pattern is distinctive to the crystal structure.

Hydrophilic—Attracted to water or polar regions of molecules.

trolled mixing, heating, and cooling programs that generate highly reproducible gelatinization and pasting profiles. The arbitrary units of measurement are the Brabender Unit (BU) and the Rapid Visco Unit (RVU) for the viscoamylograph and RVA, respectively. A third instrument, the Ottawa Starch Viscometer, first described in 1977 (8), has a stronger mixing action than the amylograph, uses smaller sample sizes, and has variable bowl speed and electronic recording of results. However, its method of heating, a boiling water bath, has a number of disadvantages, and it is slower than the RVA.

Although the RVA is becoming increasingly popular, the historical standard in the food industry is the viscoamylograph. A pressurized version, with which the effect of pressure on the cooking process can be profiled, is also available. This instrument is often used to characterize the starches used in the canning industry. Since most commercial starches are still marketed on the basis of viscoamylograph specifications, let us first review that instrument.

As a starch slurry is heated in a viscoamylograph, typically under a constant rate of shear, the increase in viscosity is measured as torque on the spindle, and a curve is traced. To a certain extent, the viscosity profile can be thought of as a reflection of the granular changes

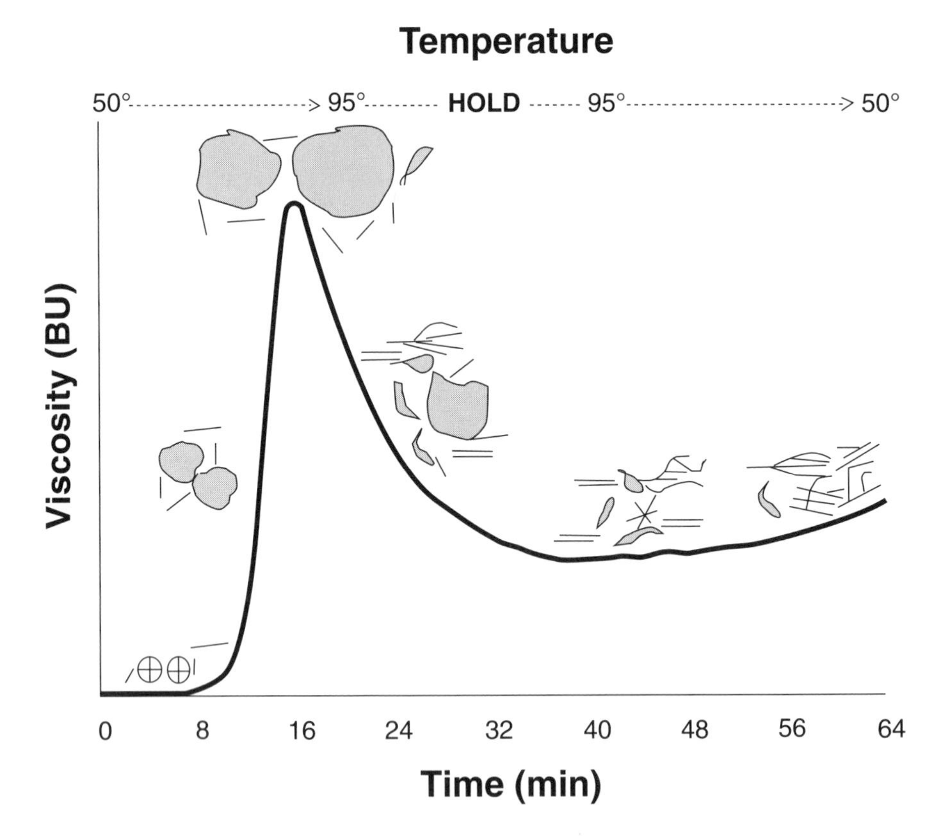

Fig. 2-9. A schematic representation of granular changes in relationship to viscosity. The viscosity profile was measured by a viscoamylograph.

that occur during gelatinization. The key point to remember is that different starches generate different viscosity profiles. An example viscoamylograph curve for native waxy corn starch and a sketch of the granular morphology during cooking are shown in Figure 2-9. During the initial heating phase, a rise in viscosity is recorded as granules begin to swell. At this point, polymers with lower molecular weights, particularly amylose molecules, begin to leach from the granules. A peak viscosity is obtained during pasting when there is a majority of fully swollen, intact granules and molecular alignment of any solubilized polymers has not yet occurred within the shear field of the instrument. During the high-temperature hold phase (e.g., 95°C [203°F]), the granules begin to break down, polymer solubilization continues, and molecular alignment occurs within the shear field of the instrument; i.e., the spindle begins to "cut the same path" through the paste. At this point, a drop in viscosity is recorded. During the cooling phase, solubilized amylose and amylopectin polymers begin to reassociate, and another rise in viscosity is traced. This second rise in viscosity is usually referred to as *set-back*. In general, for those starches that can be easily gelatinized and pasted in a viscoamylograph, the more amylose they contain, the more dramatic the set-back. Depending upon the type of starch (e.g., the botanical source and whether it is a native starch or has been chemically modified), the solids level, the pH of the slurry, and the heating regime, various gelatinization or pasting profiles can be generated.

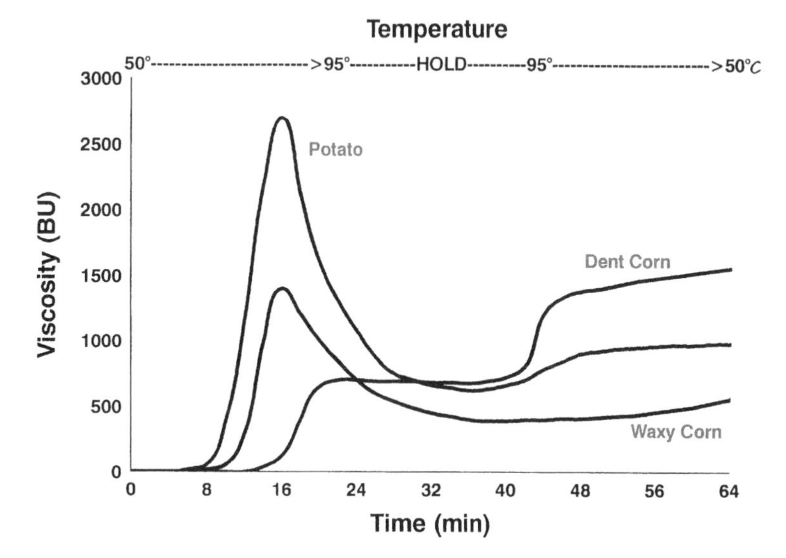

Fig. 2-10. Viscosity profiles (i.e., viscoamylograph traces) of potato, dent (common) corn, and waxy corn starch.

By keeping the viscoamylograph procedure constant, relative differences between starches can be assessed. The viscosity profiles of three major types of native starch are shown in Figure 2-10. When these profiles are plotted on the same graph, it is easy to detect viscosity differences among these starches. For example, the high peak viscosity of potato starch achieved during the initial heating phase and the ability of dent corn to set-back during the cooling cycle are both obvious from this plot. Given the number of possible cooking conditions that can be used to gelatinize starch, a wide variety of viscoamylograms can be generated by using different heating and cooling programs. Viscosity profiling is extremely helpful in determining starch behavior under various conditions and for comparing relative differences between starches. One of the most important aspects of viscosity profiling is the measurement of the effects of starch-modifying reagents or processes on gelatinization and pasting. For example, the

Set-back—Generally, the reassociation of solubilized starch polymers and insoluble granular fragments during the cooling phase of a viscometric analysis.

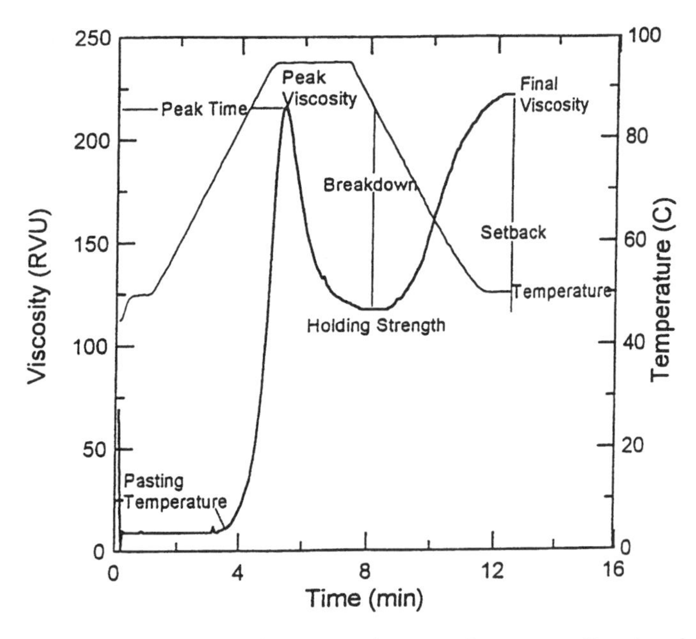

Fig. 2-11. Typical Rapid Visco-Analyser pasting curve. (Reprinted from Zhou, M., Robards, K., Glennie-Holmes, M., and Helliwell, S., 1998, Structure and pasting properties of oat starch, Cereal Chem. 75:273-281)

normally narrow peak viscosity range of native waxy corn starch (Fig. 2-10) can be stabilized over a much longer period of time by chemical modifications that prevent the granules from breaking down.

The RVA is becoming increasingly popular for the analysis of starch gelatinization and pasting. Measurements in a typical RVA curve include peak viscosity, peak area, time to peak, drop-off, and final viscosity (Fig. 2-11). Many people in the industry think that the RVA has distinct advantages over the traditional viscoamylograph: shorter analysis times (typically 20–30 min for the RVA versus 45–90 min for the viscoamylograph), smaller sample sizes, computerized control and data collection, and easier calibration procedures. The viscoamylograph, unlike the RVA, must be calibrated against a standard instrument.

Other Measurements

SWELLING POWER

The *swelling power* values of various starches can be determined by heating a weighed dry starch sample in water at 95°C for 30 min, isolating the swollen starch via centrifugation, and weighing the sediment. Swinkels (9) determined values of 21 for wheat, 24 for dent corn, 64 for waxy corn, 71 for tapioca, and 1,153 for potato. As the data reveal, the swelling power of potato is significantly higher than that of the other starches listed. The high swelling power of potato is consistent with the fact that potato starch produces a pasting curve with a very high peak viscosity. This data is also consistent with the high water-holding capacity of potato starch.

Swelling power—Swollen sediment weight (in grams) per gram of dry starch.

STARCH CONTENT

The total starch content of a complex system such as a food product can be determined by using enzymes that hydrolyze glycosidic linkages in the starch polymers. In AACC Method 76-11 (10), the enzyme glucoamylase, which reduces the amylose and amylopectin to glucose, is used. The glucose obtained is subsequently measured by a colorimetric procedure, and the total amount of starch in the sample can be calculated. In AACC Method 76-12 (10), used for starch determinations of cereal grains and products, a variety of enzymes are used to reduce the starch to glucose before a similar colorimentric determination can be made.

DAMAGED STARCH

Damaged starch is the portion of starch that is mechanically disrupted during the processes used to extract or refine starch. It is more hygroscopic than native starch. It is also susceptible to hydrolysis by the enzyme α-amylase, and this property is used to quantitate the amount of damaged starch in a sample.

AACC Method 76-30A (10) describes a procedure in which the level of damaged starch is estimated by measuring the amount of reducing groups produced by the action of α-amylase on a starch sample. In AACC Method 76-31 (10), glucoamylase is used after hydrolysis by α-amylase to produce glucose, which is then measured colorimetrically.

AMYLOSE AND AMYLOPECTIN CONTENT

The procedure employed to quantitate the amount of amylose and amylopectin in a starch sample is based on the different abilities of these starch polymers to bind iodine. As discussed previously, compared with amylopectin, amylose has a greater affinity for iodine. A quantitative method for determining iodine affinity involves titrating a starch sample with iodine and measuring the suspension with a potentiometer (11). A pure amylose sample binds about 18–19% iodine by weight. Amylopectin binds very little iodine. Values for amylopectin are generally <0.5%.

GELATINIZATION

When a native starch granule is gelatinized, several events take place. As described in Chapter 3, "gelatinization" is a broad term applied to all the changes that take place during this process. When measuring gelatinization, therefore, it is very important to be cognizant of the principle behind the method employed. What is really being measured is one of the specific events that occurs as the starch gelatinizes.

Perhaps the most widely used means of quantitating gelatinization is the polarized microscopic evaluation of birefringence described previously in the microscopy section of this chapter (12). Light-scattering techniques, based on the general principle that highly ordered structures yield intense scattering patterns under polarized light, have also been developed (13). Standard light microscopy, in conjunction with certain stains, has also been used to quantitate gelatinization and is based on the observation that native granules do not accept certain stains as readily as gelatinized granules (14).

Native starch granules are highly resistant to the action of most enzymes. When the starch gelatinizes, however, it becomes highly susceptible to the hydrolytic actions of the amylases. In one procedure, the starch-hydrolyzing enzyme glucoamylase is used (15).

Differential scanning calorimetry and related techniques, which monitor thermal events (16), and X-ray diffraction (17), as described previously, have been applied to the quantitation of gelatinization.

Damaged starch—Starch that has been mechanically disrupted by shearing during processing.

Finally, nuclear magnetic resonance, which reflects changes in molecular mobility, has been used to evaluate gelatinization (18).

References

1. Pyler, E. J. 1988. The carbohydrates. Pages 1-45 in: *Baking Science and Technology,* 3rd ed., Vol. 1. Sosland, Merriam, KS.
2. Anonymous. Microscopic examination of cooked starch. In: Pure Food Starches. Tech. Bull. Grain Processing Corporation, Muscatine, IA.
3. Jane, J., Kasemsuwan, T., Leas, S., Zobel, H., and Robyt, J. F. 1994. Anthology of starch granule morphology by scanning electron microscopy. Starch/Staerke 46:121-129.
4. Bowler, P., Williams, M. R., and Angold, R. E. 1980. A hypothesis for the morphological changes which occur on heating lenticular wheat starch in water. Starch/Staerke 32:186-189.
5. Hoseney, R. C., Atwell, W. A., and Lineback, D. R. 1977. Scanning electron microscopy of starch isolated from baked products. Cereal Foods World 22:56-60.
6. Hoseney, R. C. 1994. *Principles of Cereal Science and Technology,* 2nd ed. American Association of Cereal Chemists, St. Paul, MN.
7. Zobel, H. F. 1992. Starch granule structure. Pages 1-36 in: *Developments in Carbohydrate Chemistry.* R. J. Alexander and H. F. Zobel, Eds. American Association of Cereal Chemists, St. Paul, MN.
8. Voisey, P. W., Paton, D., and Timbers, G. E. 1977. The Ottawa Starch Viscometer—A new instrument for research and quality control applications. Cereal Chem. 54:534-557.
9. Swinkels, J. J. M. 1985. Composition and properties of commercial native starches. Starch/Staerke 37:1-5.
10. American Association of Cereal Chemists. 1995. Approved Methods of the AACC, 9th ed. Method 76-11, approved October 1976, reviewed October 1982 and October 1994; Method 76-12, approved October 1993, reviewed October 1994; Method 76-30A, approved May 1969, revised November 1972, October 1982, and October 1984, reviewed October 1994; Method 76-31, approved January 1995. The Association: St. Paul, MN.
11. Schoch, T. J. 1964. Iodimetric determination of amylose. Potentiometric titration, standard method. Methods Carbohydr. Chem. 4:157-160.
12. Watson, S. A. 1964. Determination of starch gelatinization temperature. Methods Carbohydr. Chem. 4:240-241.
13. Marchant, J. L., and Blanshard, J. V. 1978. Studies of the dynamics of the gelatinization of starch granules employing a small angle light scattering system. Starch/Staerke 30:257-264.
14. Jones, C. R. 1940. The production of mechanically damaged starch in milling as a governing factor in the diastatic activity of flour. Cereal Chem. 17:133-169.
15. Shetty, R. M., Lineback, D. R., and Seib, P. A. 1974. Determining the degree of starch gelatinization. Cereal Chem. 51:364-375.
16. Ghiasi, K., Hoseney, R. C., and Varriano-Marston, E. 1982. Gelatinization of wheat starch. III. Comparison by differential scanning calorimetry and light microscopy. Cereal Chem. 59:258-262.
17. Nara, S., Mori, A., and Komoya, T. 1978. Study on relative crystallinity of moist potato starch. Starch/Staerke 30:111-114.
18. Jaska, E. 1971. Starch gelatinization as detected by proton magnetic resonance. Cereal Chem. 48:437-444.

CHAPTER
3

Gelatinization, Pasting, and Retrogradation

Native starch granules are essentially insoluble in cold water. Thus, the unique characteristics of many of our foods, from the mouth-feel of gravies to the texture of gum drops and pie fillings, are the results not of inherent properties or behaviors of native granules, but of the changes they undergo when they are heated with water.

The first of these changes are gelatinization and pasting. They are irreversible and dependent in complex ways on the amount of heat and water available to the system. These changes render all or part of the material in granules soluble and consequently able to contribute to food properties such as texture, viscosity, and moisture retention. The third change, retrogradation, involves reassociation of the molecules and occurs after heating. The rate and extent of retrogradation are also dependent on temperature but in a manner different from that of gelatinization and pasting.

This discussion of gelatinization, pasting, and retrogradation focuses solely on granular *native starch*. It is extremely important to keep in mind that the modification of native starches, via chemical, physical, and/or enzymatic treatment, can dramatically alter their properties—usually making them better suited for food applications. In fact, because of the limitations inherent in native starches, most food products are formulated with their modified counterparts. The modification of starch is discussed in Chapter 4.

Depending upon the botanical source, native starches exhibit different behavior during pasting; i.e., no two types of starches are alike. Because so-called *cook-up starch* is typically used in food systems at about 2–6% by weight of the final product, the following discussion is based on characteristics at or near this concentration range.

In This Chapter:

Gelatinization

Researchers have tried to accurately define starch gelatinization for many years. Some claim that gelatinization starts when granular birefringence is lost; others believe it begins when an increase in viscosity occurs. Words such as "cooking," "solubilizing," and "thickening," although not exactly accurate from a starch chemist's point of view, are quite often used to characterize the gelatinization process.

Perhaps the most comprehensive definition of gelatinization was that generated from a survey of scientists and technologists taken at

Native starch—Any granular starch that has been isolated from the original plant source but has not undergone subsequent modification, i.e., unmodified starch.

Cook-up starch—Any granular starch that requires water and heat in order to gelatinize and paste.

the Starch Science and Technology Conference in 1988 (1). On the basis of the results of this survey, the following definition of gelatinization was proposed:

> Starch gelatinization is the collapse (disruption) of molecular orders within the starch granule manifested in irreversible changes in properties such as granular swelling, native crystalline melting, loss of birefringence, and starch solubilization. The point of initial gelatinization and the range over which it occurs is governed by starch concentration, method of observation, granular type, and heterogenities within the granule population under observation.

Although this definition was an attempt to clarify terms so that researchers in the field could communicate clearly, it probably will not satisfy everyone. The critical points to make here are that gelatinization

- is the disruption of molecular order;
- is temperature and moisture dependent;
- is irreversible;
- initially increases the size of granules (i.e., causes granular swelling);
- results in increased solution or suspension viscosity;
- differs with respect to cooking conditions (e.g., pH and solids);
- and differs with respect to granule type (botanical source).

Granular starch is essentially insoluble in cold water, and even when it is added to water at room temperature, little happens until heat is applied. A combination of heat and water, however, causes uncooked granules to undergo unique and irreversible changes, the most dramatic of which are 1) the disruption of the semicrystalline structure, as evidenced by a loss of birefringence; and 2) an increase in granule size, although not all granules within a given population swell at the same rate or to the same extent (Fig. 3-1). As these changes are taking place, there is an attendant increase in the viscos-

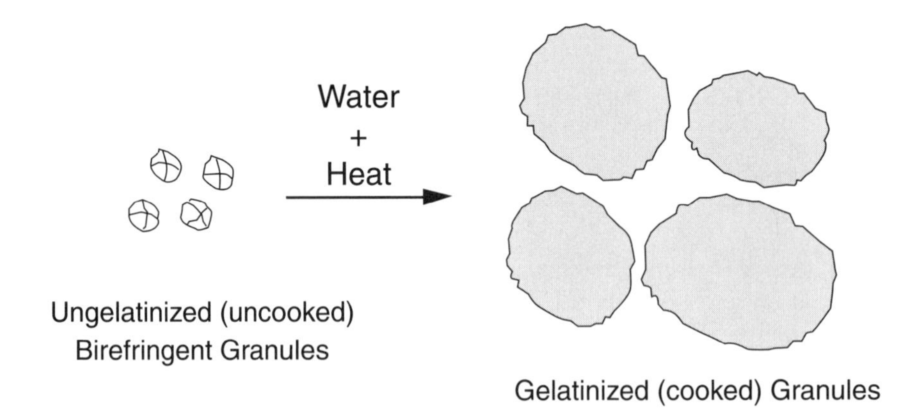

Fig. 3-1. Granular swelling. Upon heating, raw starch imbibes water and swells, resulting in an increase in paste viscosity.

ity of the medium in which the starch is heated. When a majority of the granules have undergone this process, the starch is considered to be "pasted" or "cooked-out." In most cases, it is this pasting (i.e., viscosity-forming) ability that makes starch so functional as a food ingredient.

The temperature at which starch begins to undergo these changes is referred to as the *gelatinization temperature*. In actuality, because not all the granules of a given starch begin to gelatinize at the exact same temperature, the gelatinization temperature is more appropriately defined as a relatively narrow temperature range rather than one specific temperature. Temperatures also vary depending on the source of the starch. In general, the gelatinization temperature of tuber and root starches such as potato and tapioca is slightly lower than that of cereal starches such as corn and wheat.

Gelatinization temperatures (°C) of common food starches: dent corn, 62–80; waxy corn, 63–72; wheat, 52–85; potato, 58–65; and tapioca, 52–65 (2).

From a mechanistic perspective, heating of starch in water causes disruption of the hydrogen (H) bonds between polymer chains, thereby weakening the granule. It is believed that the initial swelling takes place in the amorphous regions of the granule where hydrogen bonds are less numerous and the polymers are more susceptible to dissolution. As the structure begins to weaken, the granule imbibes water and swells. Because not all the granules gelatinize simultaneously, different degrees of structural disruption and swelling exist.

Pasting

As heating continues and more and more granules become swollen, the viscosity of the medium increases. The paste viscosity reaches a maximum when the largest percentage of swollen, intact granules is present; this is referred to as the *peak viscosity*. At this point, the starch is considered to be fully pasted. For native starches, continued heating eventually results in a decrease in viscosity as granules dissolve and polymers are solubilized.

The following proposed definition of the pasting process differentiates it from gelatinization (1):

> Pasting is the phenomenon following gelatinization in the dissolution of starch. It involves granular swelling, exudation of molecular components from the granule, and eventually, total disruption of the granules.

It is important to note, however, that pasting is not exactly separate from gelatinization, but rather an overlapping occurrence perhaps best described as a continuation of gelatinization. Although there is no definitive point at which gelatinization ends and pasting begins, pasting is usually related to the development of viscosity. However, this link between gelatinization and pasting often results in these two terms being used interchangeably to describe the same process.

When starch granules are heated in water, the smaller, linear amylose molecules begin to solubilize and leach from the granules as they

Gelatinization temperature—A narrow temperature range at which starch granules begin to swell, lose crystallinity, and viscosify the cooking medium.

Peak viscosity—The point at which, during heating in water, gelatinized starch reaches its maximum viscosity.

swell. As heating continues, additional amylose as well as amylopectin are "released" from the granule. Paste viscosity is highest when a majority of fully swollen, intact granules are present in the cooking medium. If heating continues, solubilization of polymers increases as granules begin to break down and structural integrity is lost.

For most starches, the texture and viscosity of the resultant paste change when it is cooled. Either a *viscoelastic paste* or a *gel* is usually formed, depending upon the amylose content. In general, the higher the amylose content, the more likely the paste will set to a firm and cuttable gel. A firm gel is most often the result of noncovalent reassociation of the linear amylose molecules after heating. The amylose and amylopectin content of a particular starch dictate the gelatinization behavior and rheological characteristics of the resultant paste or gel. For example, waxy corn starch (containing essentially all amylopectin) has an initially high peak viscosity followed by viscosity breakdown during continued heating. After cooling to either ambient or refrigeration temperature, the resultant paste is considered to be cohesive and viscoelastic in nature but does not set to a gel. On the other hand, dent corn starch, which contains approximately 25% amylose, shows a relatively low peak viscosity and limited viscosity breakdown during heating. After cooling, however, dent corn starch usually sets to a cuttable gel.

Retrogradation

Solubilized starch polymers and the remaining insoluble granular fragments have a tendency to reassociate after heating. This reassociation is referred to as retrogradation. One definition of starch retrogradation is as follows (1):

> Starch retrogradation is a process which occurs when starch chains begin to reassociate in an ordered structure. In its initial phases, two or more starch chains may form a simple juncture point which then may develop into more extensively ordered regions. Ultimately, under favorable conditions, a crystalline order appears.

Retrogradation is especially evident when amylose-containing starches are cooled. Upon cooling, less energy is available to keep the solubilized starch molecules apart. Retrogradation results in the formation of crystalline aggregates and a gelled texture. Although amylopectin can retrograde upon cooling, linear amylose molecules have a greater tendency to reassociate and form hydrogen bonds than the larger, "tumbleweed-like" amylopectin molecules. As the retrogradation process occurs, the starch paste becomes increasingly opaque and forms a cuttable gel. With time, this gel becomes rubbery and has a tendency to release water. The release of water is referred to as *syneresis* and is commonly found in food products such as sauces and dips that are formulated with unmodified amylose-containing starch. Because the retrogradation process is amplified when heated starch is

Viscoelastic paste—A paste that possesses properties of both a viscous (liquid-like) and an elastic (solid-like) substance.

Gel—A liquid system that has the properties of a solid.

Syneresis—The separation of a liquid from a gel; weeping.

cooled, syneresis is most prevalent in refrigerated and frozen products. Freeze-thaw stability, measured by the degree of syneresis, or water released, is an important consideration for food starches and should be kept in mind when refrigerated and frozen foods are formulated. It is also important to remember that starches from different botanical sources retrograde at different rates and to various degrees. For example, tapioca starch with about 19% amylose sets to a soft gel over time, whereas high-amylose corn starch (i.e., starch obtained from a hybrid corn plant that contains greater than about 40% amylose) sets to a very firm gel relatively quickly after being heated and cooled.

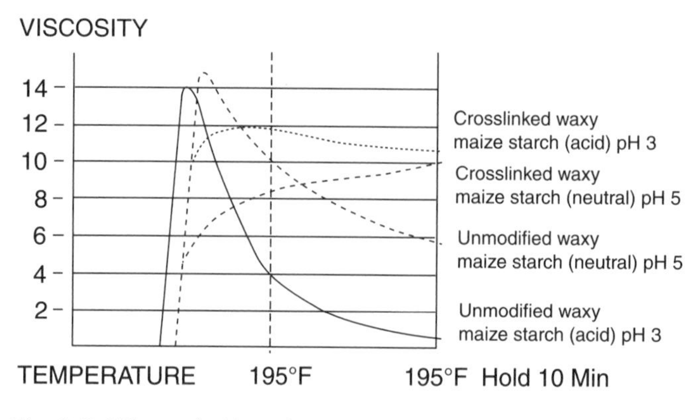

Fig. 3-2. Effects of pH on the pasting profiles of native and chemically modified waxy corn. (Source: Smith, P. H., 1982, p. 251 in: *Food Carbohydrates,* D. R. Lineback and G. E. Inglett, eds., AVI, Westport, CT)

Effects of pH, Shear, and Other Ingredients

Since foods can vary dramatically in pH, the effect of pH on starch gelatinization is an important consideration. As shown in Figure 3-2, viscosity profiles can vary significantly as a function of pH, not only for native starch but for modified (e.g., crosslinked) starch as well. In general, pH extremes tend to have a negative impact on viscosity by hydrolyzing bonds and disrupting the molecular integrity of the granule. Although extreme pH environments can actually help to gelatinize the starch during early stages of heating, the starch soon begins to break down. In particular, acid hydrolysis can lead to the degradation of starch into products with lower molecular weights, especially during protracted heating. The net result of acid hydrolysis is viscosity loss.

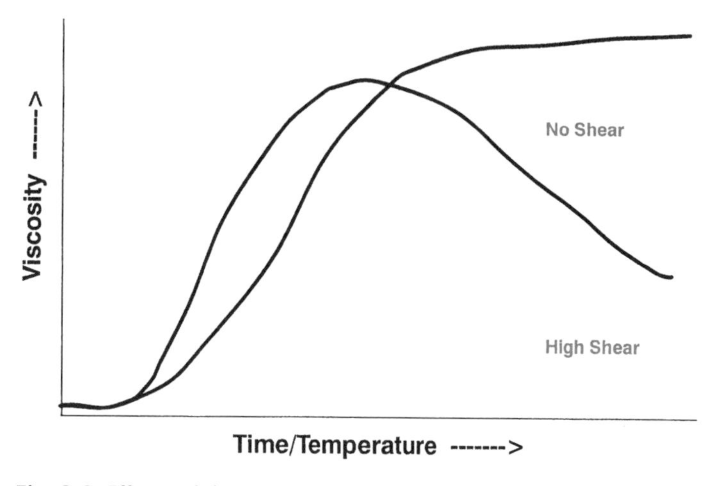

Fig. 3-3. Effects of shear on the pasting profile of unmodified starch.

Shear can also have a dramatic impact on starch behavior. As shown in Figure 3-3, a combination of heating and shearing can have a drastic effect on starch gelatinization and paste viscosity. Unmodified starches, which are normally susceptible to granular disruption and viscosity breakdown, are even more vulnerable when subjected to shear during the gelatinization process. Shear can also have detrimental effects on paste viscosity after gelatinization, again particularly on native starch (Fig. 3-4). Shear can come from various types of processing equipment such as high-speed mixers, pumps,

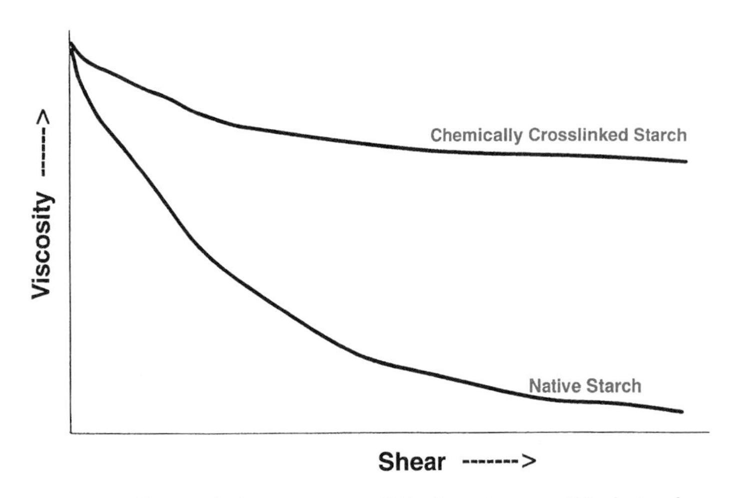

Fig. 3-4. Effects of shear on crosslinked and unmodified starches after gelatinization.

and homogenizers. Whenever starch is subjected to processing conditions, there is a possibility for increased granular breakdown and loss of viscosity.

Because food products are usually quite complex, the effect of other ingredients also has to be considered when evaluating starch functionality in terms of viscosity. Fats, sugars, proteins, and salts can all influence starch gelatinization, pasting, and retrogradation. In general, any ingredient that either interacts (e.g., coats, binds, or complexes) with the granule or competes with the granule for available water can have a negative impact on viscosity. For example, fat has a tendency to interact with the starch granule and prevent complete hydration, resulting in lower viscosity development. Sugar and other solids limit gelatinization and pasting by competing for available water. Other food ingredients, such as proteins and salts, can also alter starch performance and should be considered when starch-containing foods are formulated.

References

1. Atwell, W. A., Hood, L. F., Lineback, D. R., Varriano-Marston, E., and Zobel, H. F. 1988. The terminology and methodology associated with basic starch phenomena. Cereal Foods World 33:306-311.
2. Whistler, R. L., and BeMiller, J. N. 1997. *Carbohydrate Chemistry for Food Scientists*. American Association of Cereal Chemists, St. Paul, MN.

CHAPTER
4

Starch Modifications

In This Chapter:

Native starches from various plant sources have their own unique properties. To the extent possible, these inherent characteristics are exploited by food processors to meet specific needs. Native starches, however, lack the versatility to function adequately in the entire range of food products currently available in the marketplace. The diversity of the modern food industry and the enormous variety of food products require that starch be able to tolerate a wide range of processing techniques as well as various distribution, storage, and final preparation conditions. These demands are met by modifying native starches by chemical and physical methods. This chapter highlights the technologies commonly practiced by the starch manufacturer for the production of modified starches for the food industry. For additional information, more comprehensive reviews are available (1,2).

Chemical Modification

The most common type of starch modification is the treatment of native starch with small amounts of approved chemical reagents. Chemical modification of starch changes the functionality of the starch. The chemistry involved in the modification of starch is quite straightforward and involves primarily reactions associated with the hydroxyl groups of the starch polymer. Derivatization via ether or ester formation, oxidation of the hydroxyl groups to carbonyl or carboxylic groups, and hydrolysis of glycosidic bonds are some of the major mechanisms of chemical modification. The chemical reagents discussed in this chapter have all been approved by regulatory agencies in the United States for producing modified starches for food use. Allowable reagents and usage levels for the production of "Food Starch—Modified" are defined in Title 21 of the Code of Federal Regulations (Section 172.892).

GENERAL CONSIDERATIONS

The chemical modification of starch is usually performed in an aqueous medium. A suspension of starch in water, typically 30–45% solids (by weight), is treated with the chemical reagent(s) under proper agitation, temperature, and pH. When the reaction is complete, the starch is brought to the desired pH by a neutralizing agent

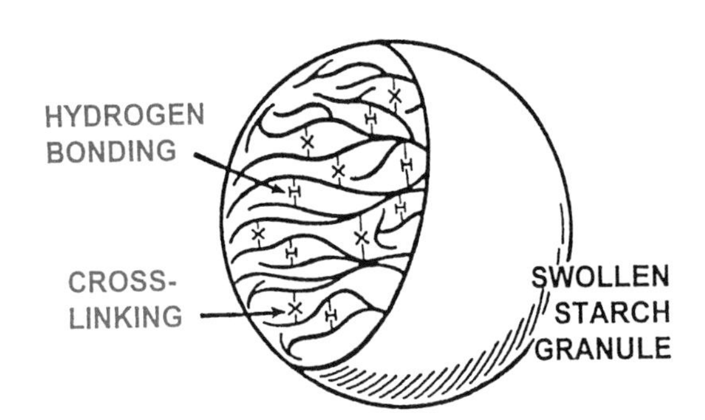

Fig. 4-1. Structure of a crosslinked starch granule. Each X refers to a covalently bonded crosslinking reagent.

and then purified by washing with water and recovered as a dry powder. In most cases, the derivatization efficiency is about 70% or more.

The extent of chemical modification is generally expressed as the *degree of substitution* (*DS*) when the substituent group (e.g., acetate or phosphate) reacts with the hydroxyl groups of the D-glucopyranosyl unit. *Molar substitution* (*MS*) is used when the substituent group can further react with the reagent itself to form a polymeric substituent. The average mole number of substituent groups per D-glucopyranosyl unit is determined by chemical or physical methods, depending on the nature of the substituent. Simple weight percentages are also used to indicate the amount of modification found in starch. Nuclear magnetic resonance is a useful analytical tool used to identify the degree of substitution and specific position of the substituent group on the glucose molecule within the starch polymer. Significant amounts of commercial starch are modified by crosslinking or substitution reactions, which can be done either separately or in combination.

CROSSLINKING

Perhaps the most common type of chemical modification technique is *crosslinking,* the derivatization of starch by using a bi- or polyfunctional chemical reagent that is able to react with two or more different hydroxyl groups on the same or different starch polymers. Covalently bonded crosslinks act as "spot welds" to reinforce the granular structure (Fig. 4-1). Crosslinking controls granular swelling and produces a starch that can tolerate high temperature, high shear, and acidic conditions.

Crosslinked starch is typically manufactured by reacting an alkaline slurry (pH 7.5–12) of granular starch (30–45% solids) with an approved crosslinking reagent (often in the presence of salt). Among the chemicals approved and used for crosslinking starch are phosphorus oxychloride, sodium trimetaphosphate, and mixes of adipic anhyhdride and acetic anhydride. The temperature of the reaction is typically 25–50°C (77–122°F), and reaction times can vary from 30 min to 24 hr, depending upon the protocol. After the reaction is complete, the starch is adjusted to neutral pH, filtered, washed, and dried.

Phosphorus oxychloride. Phosphorus oxychloride is commonly used to produce crosslinked starch esters. The reaction is typically run at a high pH (approximately 11.5) with or without salts. Salts commonly used include sodium chloride and sodium sulfate. The reaction, as shown in Equation 4-1, is usually complete within an hour at room temperature. Depending upon the extent of crosslinking, the di-starch phosphate that is generated produces a starch with improved viscosity stability and process tolerance.

Degree of substitution (DS)—Measurement of the average number of hydroxyl groups on each D-glucopyranosyl unit (commonly called an anhydroglucose unit [AGU]) that are derivatized by substituent groups. Since the majority of AGUs in starch have three hydroxyl groups available for substitution, the maximum possible DS is 3.

Molar substitution (MS)—Level of substitution in terms of moles of monomeric units (in the polymeric substituent) per mole of AGU. The MS can be greater than 3, given the substituent's ability to react further.

Crosslinking—Chemical modification of starch that results in covalently bonded inter- and intramolecular bridges between starch polymers.

Sodium trimetaphosphate. Like that of the phosphorus oxychloride treatment, the product of this crosslinking reaction is a di-starch phosphate (ester linkage). This reaction (Equation 4-2), however, requires a much longer preparation time because of the nature of the reagent. At least several hours are required versus the short reaction time for phosphorus oxychloride. The product possesses properties similar to those of phosphorus oxychloride crosslinked starch.

$$\underset{\text{Cl}\;\;\text{Cl}\;\;\text{Cl}}{\overset{\text{O}\,\|}{\text{P}}} + \text{StOH} \xrightarrow{\text{NaOH}} \underset{|\;\text{ONa}}{\overset{\text{O}\,\|}{\text{StOPOSt}}} + \text{NaCl}$$

Equation 4-1. Chemical reaction for phosphorus oxychloride crosslinking of starch. St = starch polymer.

$$2\text{StOH} + Na_3P_3O_9 \xrightarrow[\text{CATALYST}]{\text{ALKALI}} \underset{|\;\text{ONa}}{\overset{\text{O}\,\|}{\text{StOPOSt}}} + Na_2H_2P_2O_7$$

Equation 4-2. Chemical reaction for sodium trimetaphosphate crosslinking of starch. St = starch polymer.

Mixed adipic and acetic anhydride. The mixed anhydride reagent used for this reaction (Equation 4-3) creates organic ester linkages in starch that are relatively stable under conditions of neutral pH. Under conditions of extreme pH, however, these types of crosslinks are less stable than ether or inorganic ester linkages. Therefore, the mixed anhydride reaction is conducted at a pH of less than 9. When the mixed anhydride reagent is used, the starch is also substituted with acetyl groups.

Effects of crosslinking. Crosslinking can have a dramatic effect on the viscosity profile of starch (Fig. 4-2). Starch, which is normally susceptible to viscosity breakdown either from prolonged heating, high shear, acidic conditions, or all three, shows a stable viscosity profile over time once it is crosslinked. Crosslinked starch is sometimes referred to as "inhibited" starch because crosslinking tends to inhibit granular swelling during cooking. Starch that is lightly crosslinked (low-DS starch), however, tends to show a peak viscosity that is actually higher than that of its native unmodified starch. The key benefits of crosslinking, even at low levels, are granular stability and improved paste texture. For example, the normally cohesive, gummy consistency associated with native waxy corn starch is eliminated, and a short, salve-like texture is produced. In general, as the level of crosslinking increases, the starch becomes more resistant to the changes generally associated with cooking and pasting. Starch with a high level of crosslinking (high-DS starch) does not have a peak vis-

$$\text{StOH} + CH_3\overset{\text{O}}{\overset{\|}{\text{C}}}\text{O}\overset{\text{O}}{\overset{\|}{\text{C}}}(CH_2)_n\overset{\text{O}}{\overset{\|}{\text{C}}}\text{O}\overset{\text{O}}{\overset{\|}{\text{C}}}CH_3 \xrightarrow[\text{pH 8}]{\text{NaOH}} \text{StO}\overset{\text{O}}{\overset{\|}{\text{C}}}(CH_2)_n\overset{\text{O}}{\overset{\|}{\text{C}}}\text{OSt} + CH_3\overset{\text{O}}{\overset{\|}{\text{C}}}\text{ONa}$$

Equation 4-3. Chemical reaction for crosslinking of starch by a mixed adipic-acetic anhydride reagent. St = starch polymer.

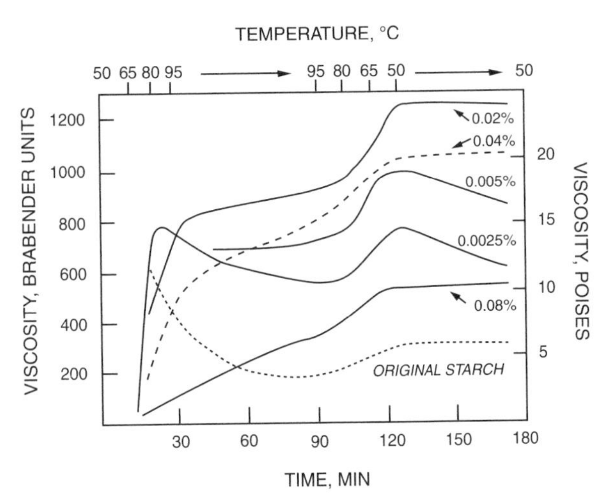

Fig. 4-2. Effect of crosslinking (% trimetaphosphate treatment) on the viscosity profile of starch. (Source: Smith, P. H., 1982, p. 250 in: *Food Carbohydrates,* D. R. Lineback and G. E. Inglett, eds., AVI, Westport, CT)

cosity but rather shows a continual and progressive increase in viscosity throughout the cooking process. Since starches with various levels of crosslinking are commercially available, it is important to select the level that allows for optimal performance in a particular system. For example, a starch that is highly crosslinked (i.e., highly inhibited) may not function appropriately in a system that receives a limited amount of heat. The objective is to select a level of crosslinking that protects the starch during processing, especially under conditions of high shear or acid cooking, but allows for adequate granular swelling and viscosity formation.

SUBSTITUTION (STABILIZATION)

As discussed in previous chapters, the viscosity of starch pastes tends to increase upon cooling and aging. This viscosity formation and textural change can be extreme for certain types of amylose-containing starches. In fact, these starches can form rigid gels caused by the reassociation of amylose molecules. This reassociation of starch polymers is usually referred to as retrogradation (see Chapter 3). In order to minimize or prevent retrogradation, starch is substituted or stabilized by introducing monofunctional chemical "blocking groups," such as acetyl or hydroxypropyl groups, along the polymer backbone (Fig. 4-3). *Substitution* lowers the gelatinization temperature and stabilizes the starch by preventing the reassociation, i.e., retrogradation, of polymers after cooking. Substituted starch is particularly useful for refrigerated and frozen food applications.

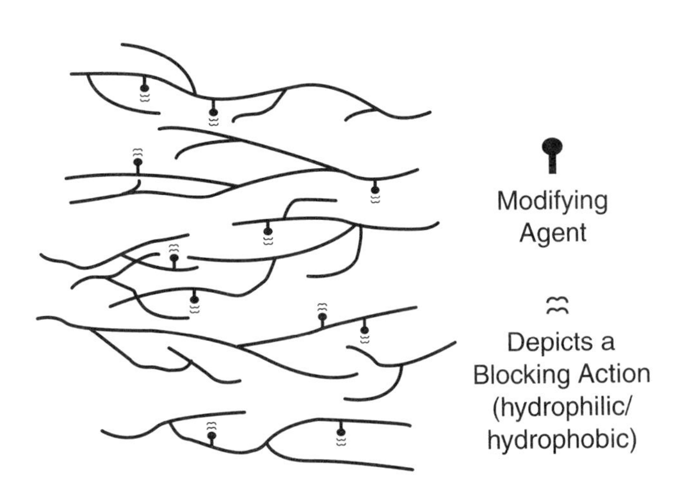

Fig. 4-3. Action of blocking groups in substituted (stabilized) starch.

Substitution—Chemical modification of starch resulting in the addition of a chemical blocking group between starch polymers and involving derivatization with a monofunctional reagent through ester or ether formation.

Starch esters. *Starch acetate.* Acetylation is one of the most common methods of stabilization in commercial practice. Starch acetate containing 0.5–2.5% acetyl groups is typically used for food applications (2). The FDA limit for food starches is 2.5% acetyl content. Acetylated starch is typically prepared by the slow addition of acetic anhydride or vinyl acetate to a starch slurry at pH 7.5–9.0. After the reaction is complete (Equation 4-4), the starch slurry is neutralized and washed. The acetylated starch is then recovered as a powder by centrifugation and drying. Acetylated starch, particularly crosslinked acetylated starch, is used primarily as a thickener because of its stability and clarity.

Acetylated starches are used in a variety of food products including refrigerated and frozen foods. When acetylated starch is used in dairy products, a curdling phenomenon, which is thought to result from the instability of the acetate linkage under high protein concentration, occasionally occurs. Commercial acetylated starches are not usually as freeze-thaw stable as hydroxypropylated starches.

$$(CH_3C(=O))_2O + StOH \xrightarrow{OH^-} CH_3\text{-}C(=O)\text{-}OSt + NaOAc$$

$$CH_3\text{-}C(=O)\text{-}O\text{-}CH=CH_2 + StOH \xrightarrow{OH^-} CH_3\text{-}C(=O)\text{-}O\text{-}St + CH_3CHO$$

Equation 4-4. Chemical reactions for acetate substitution of starch. St = starch polymer.

Starch octenylsuccinate. For most food applications, the degree of monosubstitution on the starch polymer is typically low, and there is generally little impact on the overall hydrophilic property of the starch polymer. However, when starch polymers are substituted with 1-octenylsuccinic anhydride (OSA), or succinic anhydride, a hydrophobic moiety is introduced into the polymer and a new class of starch products is created. These products are effective as emulsion stabilizers (see Chapter 9).

The chemistry for preparing OSA-treated starch is essentially the same as that for starch acetate, except that a hydrophobic cyclic anhydride is used and the product is a half-ester of succinic acid (Equation 4-5). The reaction is usually carried out at pH 7–9 in an

$$StOH + CH_3(CH_2)_4CH=CHCH_2CH\text{-}C(=O)\text{-}O\text{-}C(=O)CH_2 \text{ (cyclic)} \xrightarrow{NaOH}$$

$$CH_2(CH_2)_4CH=CHCH_2\text{-}CH(\text{-}CH_2C(=O)\text{-}ONa)\text{-}C(=O)\text{-}OSt$$

Equation 4-5. Chemical reaction for 1-octenylsuccinic anhydride substitution of starch. St = starch polymer.

aqueous starch suspension. Good agitation of the reaction mixture is recommended because of the hydrophobic nature of the OSA reagent.

The pasting behavior of OSA-treated starch is similar to that of other monosubstituted starch derivatives. For example, the gelatinization temperature decreases with increased substitution. When amylose-containing starch is derivatized with OSA, the tendency of the cooked paste to form a gel is greatly reduced. Instead, a rather rubbery texture is produced, and its viscosity is generally higher than that of other substituted derivatives. This phenomenon may be caused by an associative effect of the hydrophobic chains.

OSA-treated starch is used as a stabilizing agent in products such as beverages and salad dressings, as a flavor-encapsulating agent, as a clouding agent, and as a processing aid. One key application of OSA-treated starch is the replacement of gum arabic in systems that require emulsion stabilization and/or encapsulation (see Chapter 9).

Starch phosphate. Starch phosphates are ester derivatives of phosphoric acid. Unlike crosslinked starch, which is formed by reacting starch with polyfunctional phosphorylating agents such as phosphorus oxychloride or sodium trimetaphosphate (as described previously), starch phosphates are produced by using monofunctional reagents such as sodium orthophosphate and sodium tripolyphosphate. With orthophosphate, starch phosphate monoesters can be conveniently produced (Equation 4-6) by a dry-heat reaction at pH 5.0–6.5 for 0.5–6.0 hr at 120–160°C (248–320°F) (2). The sodium tripolyphosphate reaction can be carried out at more moderate temperatures (100–120°C [212– 248°F]). The pH of the reaction typically starts on the alkaline side at about 8.5–9.0 and finishes at a pH of less than 7.0. The reaction is shown in Equation 4-7.

$$StOH + NaH_2PO_4/Na_2HPO_4 \longrightarrow StO-\overset{\overset{\displaystyle O}{\|}}{\underset{\underset{\displaystyle OH}{|}}{P}}-ONa$$

Equation 4-6. Chemical reaction for orthophosphate substitution of starch. St = starch polymer.

Pastes of starch phosphate derivatives usually have improved clarity, low-temperature stability, and emulsifying properties. Starch

$$StOH + O=P(ONa)[O-P(=O)(ONa)_2]_2 \longrightarrow StO-\overset{\overset{\displaystyle O}{\|}}{\underset{\underset{\displaystyle ONa}{|}}{P}}-ONa + Na_3HP_2O_7$$

Equation 4-7. Chemical reaction for sodium tripolyphosphate substitution of starch. St = starch polymer.

phosphates have been recommended for salad dressing applications as a means to improve the stability of vinegar and vegetable oil emulsions.

Starch ether. Hydroxypropylated starch has brought a new dimension to the food industry because of its improved functional characteristics compared with other types of substituted starch products. Pastes of hydroxypropylated starch have improved clarity, greater viscosity, reduced syneresis, and freeze-thaw stability. If the starch is also crosslinked, freeze-thaw stability during prolonged storage periods can be further enhanced. Crosslinked, hydroxypropylated starches are perhaps the most commonly used modified starches in the food industry. They are typically used in a wide range of food applications, including gravies, dips, sauces, fruit pie fillings, and puddings—essentially anywhere a smooth, viscous, clear thickener is necessary and freeze-thaw stability is required.

Hydroxypropylated starches are prepared by reacting a starch slurry with propylene oxide under highly alkaline conditions at a temperature of approximately 30–50°C (86–122°F) (Equation 4-8). Because hydroxypropyl groups lower the gelatinization temperature and high alkalinity can begin to cause granular swelling by itself, salts such as sodium sulfate are required as part of the reaction mixture to prevent granular swelling. Hydroxypropylated starches are also made commercially by an alcohol slurry reaction.

$$StOH + CH_2\text{-}CH\text{-}CH_3 \ (\text{epoxide, } O \text{ bridging } CH_2 \text{ and } CH) \xrightarrow{OH^-} StOCH_2CH(OH)CH_3$$

Equation 4-8. Chemical reaction for hydroxypropyl substitution of starch. St = starch polymer.

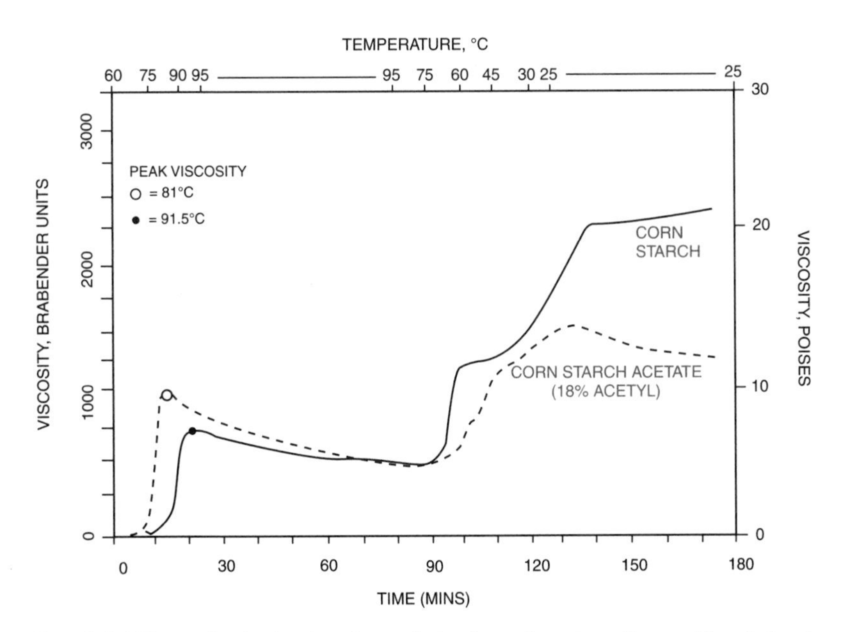

Fig. 4-4. Effect of substitution (acetylation) on the viscosity profile of starch. (Reprinted, with permission, from [2])

Effects of substitution. The effects of chemical substitution on the starch viscosity profile are shown in Figure 4-4. Substitution tends to result in 1) a lowering of the gelatinization temperature; 2) an increase in the peak viscosity; 3) a decrease in set-back during paste cooling, especially in amylose-containing starch; and 4) a stable paste that is resistant to retrogradation during storage. In simple terms, chemical substitution (i.e., the introduction of chemical blocking groups) acts to "loosen" the polymeric structure of the granule, causing it to be more susceptible to cooking. Upon cooling, however, the "blocking action" limits polymer reassociation, and retrogradation is reduced. Highly substituted starch is extremely freeze-thaw stable and has better paste clarity than its nonsubstituted native starch base. The clarity of the paste is improved by increasing the degree of swelling or hydration capacity of the starch granule and by reducing retrogradation (2).

CONVERSION

Starches normally have good thickening properties because of their high molecular weight polymeric components. For applications that typically utilize a high starch (i.e., solids) content, such as candies and food coatings, the common practice is to use starches that have been converted. The conversion process results in starch products that contain reduced molecular weight polymers and exhibit reduced viscosity.

The most common conversion methods used in the starch industry include acid hydrolysis, oxidation, pyroconversion, and enzyme conversion. Except for enzyme conversions, granular starch is used in the modification processes for ease of recovery. The properties of converted starches can vary widely depending upon the type of base starch used and the conversion process, e.g., the extent (time) of conversion and the method employed (acid, oxidant, enzyme, heat, or combinations thereof).

Acid hydrolysis. Although acid hydrolysis of starch was recorded as early as 1811 for the production of sugars and syrups from dispersed starch (3), the commercial production of acid-converted starches did not begin until about 1900.

The chemistry of the conversion process involves the acid-catalyzed hydrolysis of both α-1,4 and α-1,6 glycosidic linkages. The hydrolysis occurs preferentially in the amorphous regions of the granule, leaving starch granules with a more crystalline structure. Acid-converted starch is produced commercially by the controlled addition of acid to a starch slurry under agitation at a temperature ranging from ambient (25°C [77°F]) to a few degrees below the typical starch gelatinization temperature (e.g., 55°C [131°F]) until the appropriate degree of hydrolysis is reached. The acid is then neutralized, and the converted starch granules are recovered by filtration and then washed and dried. Hydrochloric and sulfuric acids are most commonly used in the conversion process.

Although various methods can be used to measure the extent of hydrolysis, such as molecular weight, intrinsic viscosity, Brookfield viscosity, alkaline number, and reducing value, the starch industry generally uses a simple, empirical scale referred to as *water fluidity (WF)*. The WF scale is the reciprocal of viscosity and ranges from 0 to 100, in which 0 is the viscosity of unmodified starch and 100 is the viscosity of water (Fig. 4-5). For amylose-containing starches such as dent corn, tapioca, and potato, the strongest gelling ability occurs at a WF value between 60 and 70.

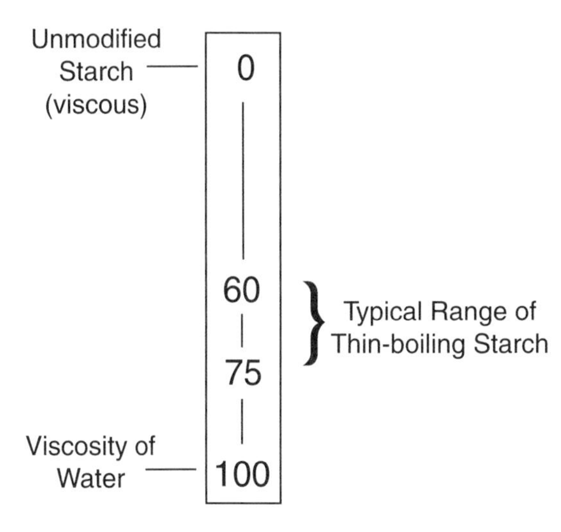

Fig. 4-5. Water fluidity (WF) scale used to measure the extent of starch conversion.

The primary objective of acid hydrolysis is to reduce the hot viscosity of the starch paste so that higher concentrations of starch can be dispersed without excessive thickening. These products are commonly referred to as "thin-boiling" or acid-thinned starches. In addition to low hot viscosity, high gel strength at high solids levels is another key attribute associated with acid-converted starches. These types of starches are particularly useful in confections where gelling is required. Candies containing these starches typically have a soft, jellylike texture that is tender yet firm. The degree of hydrolysis can be optimized to control textural properties.

Oxidation and bleaching. Food starches can be modified with oxidizing agents under controlled conditions. Depending upon the type and quantity of reagent used, this form of modification is classified as either bleaching or oxidation. Bleaching agents include hydrogen peroxide, ammonium persulfate, sodium or calcium hypochlorite, potassium permanganate, and sodium chlorite. For oxidation, only chlorine, as sodium hypochlorite, is permitted. The amount of sodium hypochlorite allowed is much higher than that used for bleaching: 0.055 versus 0.0082 pound of active chlorine per pound of starch.

The main purpose of bleaching corn starch is to improve the whiteness of the starch powder by oxidizing the impurities such as carotene, xanthophyll, and related pigments. Even though the treatment level is low, some oxidation of the hydroxyl groups occurs, especially when hypochlorite is used. Treatment levels above those used for bleaching produce converted or degraded low-viscosity products. The chemistry of hypochlorite oxidation is relatively complex but primarily involves carbons 2, 3, and 6 on a D-glucopyranosyl unit. It is generally agreed that about 25% of the oxidizing reagent is consumed in carbon-carbon splitting while about 75% oxidizes hydroxyl groups (4). The main reactions occurring during oxidation are shown in Figure 4-6.

Oxidized starch products usually have low viscosity, excellent clarity, and low-temperature stability. Oxidized starches are often used in batters and breadings for coating a wide variety of foodstuffs such as

Water fluidity (WF)—Scale used to determine the degree of starch conversion.

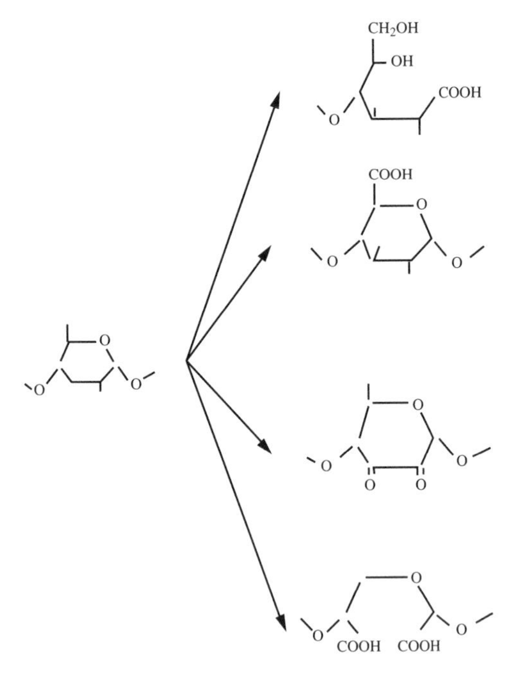

Fig. 4-6. Major reactions that occur during oxidation. (Adapted from [4])

chicken, fish, emulsion meats, and vegetables. These types of starches provide good adhesion of the batter to the food and a crispy texture after frying.

Pyroconversion (dextrinization). In pyroconversion, starch products are prepared by dry roasting acidified starch. These products are referred to as dextrins or more accurately as pyrodextrins. Depending upon the reaction conditions (e.g., pH, moisture, temperature, and length of treatment), pyroconversion produces a range of products that vary in viscosity, cold-water solubility, color, reducing sugar content, and stability.

As shown in Table 4-1, pyrodextrins are typically classified as white dextrins, yellow dextrins, and British gums, depending upon processing conditions and their resultant properties (1). Commercial pyrodextrins are generally produced by heating dry, acidified starch in a reactor with good agitation. Acid may be sprayed onto the starch to facilitate glycosidic hydrolysis. The chemistry of acid-catalyzed pyrolysis is complex. Depending upon the conditions of dextrinization, both hydrolysis and repolymerization can occur (Fig. 4-7). Pyroconversion generally creates new glycosidic bonds in addition to the existing α-1,4 and α-1,6 linkages.

Because of their typically low viscosity, good film-forming ability, and high solubility in water, pyrodextrins are used in the coating of foods and can replace more costly gums in many of these applications. Special high-viscosity dextrins can be used as fat replacers in bakery and dairy products.

Enzyme conversion. Historically, starch-based products have been hydrolyzed by fermentation processes to produce sugars, which can be further converted into alcohol. Amylases and other enzymes have been used commercially for many years to modify starch.

Many chemically modified starch products can be made with native starch granules as the starting material. However, because gelatinized starch is very susceptible to enzyme action, in most cases the

TABLE 4-1. Reaction Conditions and Properties of White Dextrin, Yellow Dextrin, and British Gum[a]

Condition	White Dextrin	Yellow Dextrin	British Gum
Roasting temperature (°C)	110–130	135–160	150–180
Roasting time (hr)	3–7	8–14	10–24
Amount of catalyst	High	Medium	Low
Solubility	Low to high	High	Low to high
Viscosity	Low to high	Low	Low to high
Color	White to cream	Buff to dark tan	Light to dark tan
Type of reaction	Mainly hydrolysis	Hydrolysis and repolymerization	Mainly repolymerization

[a] Adapted from (1), p. 596.

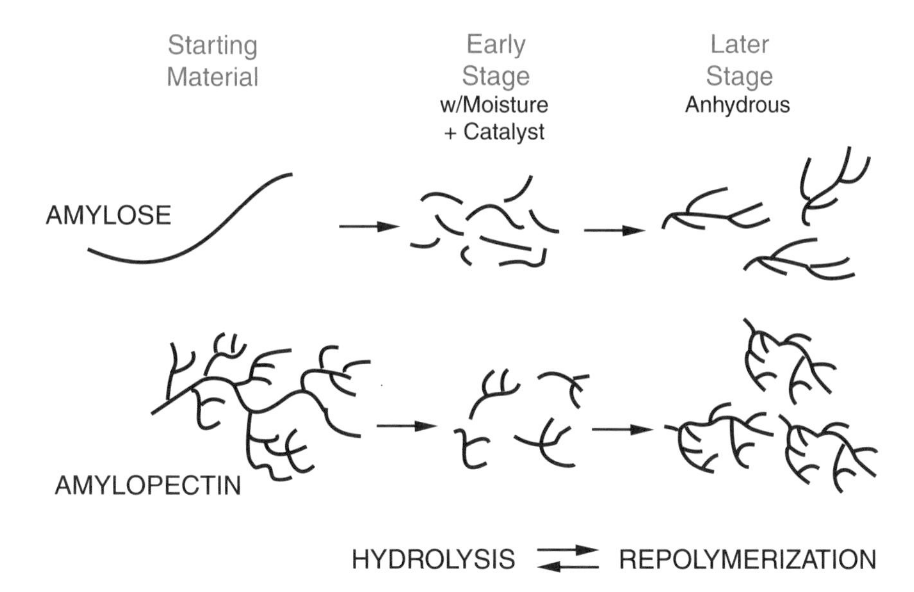

Fig. 4-7. Molecular transformations that occur during pyroconversion.

enzymatic modification of starch is carried out on a gelatinized dispersion or paste. Depending on the type and source, enzymes have different optimal pH levels and temperature conditions for maximum activity. Consequently, the pH and temperature ranges of the gelatinized starch dispersion or paste generated in industrial starch-modifying processes are controlled to maximize the efficiency of the enzyme treatment (5).

The most widely used enzymatic modification of starch is the conversion of starch to maltodextrins, corn syrups, and sugars (e.g., dextrose). There is a wide range of these products, which vary depending on the size of the molecules in the final reaction mixture. The smaller the molecules, the sweeter the mixture and the higher the *dextrose equivalent (DE)*. The DE of starch is 0, and that of dextrose is 100. Maltodextrins have a DE of less than 20 and corn syrup solids a DE of 20 or greater. *α-Amylase* is often used in the production of maltodextrins and low-DE syrups. The production of very high-DE syrups usually involves the use of *β-amylase,* glucoamylase, and/or a debranching enzyme in addition to α-amylase. The extent of starch conversion is always monitored as the reaction proceeds. Typically, the longer the conversion time, the lower the viscosity of the resultant product and the higher the reducing sugar content. Depending on the enzymes employed and the reaction conditions used, it is possible to commercially hydrolyze starch entirely to dextrose or to a variety of conversion products with molecules of intermediate sizes. In general, as the DE changes, so do functional attributes such as hygroscopicity, solubility, sweetness, viscosity, gelling, and water-binding properties. A detailed discussion of the properties of starch conversion products is beyond the scope of this book.

Dextrose equivalent (DE)—Indication of the reducing sugar content calculated as the percent anhydrous dextrose of the total dry substance. Pure dextrose has a DE of 100.

α-Amylase (α-1,4-glucan-4-glucanohydrolase)—Enzyme that acts to degrade starch polymers at an internal site anywhere in an amylose or amylopectin molecule by hydrolyzing α-1,4 linkages.

β-Amylase (α-1,4-glucan maltohydrolase)—Enzyme that cleaves alternate glycosidic bonds of starch in α-1,4 chains in a stepwise fashion starting at the nonreducing end.

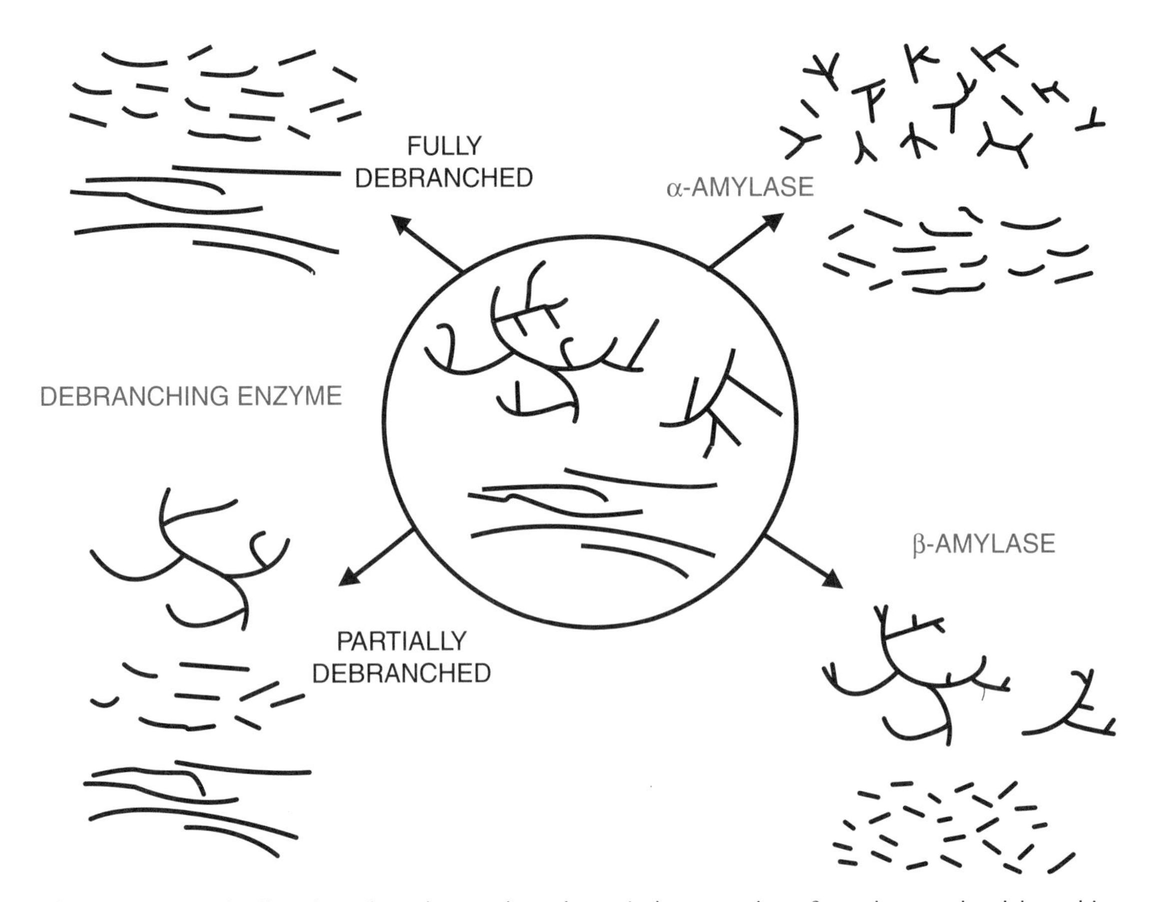

Fig. 4-8. Enzymatic digestion of amylose and amylopectin by α-amylase, β-amylase, and a debranching enzyme.

Starch conversion products can be provided in either the dry form or as syrups. The removal of water from syrups to form dry powders is not easily accomplished. Because heating can cause discoloration of high-DE syrups, vacuum is often used in conjunction with heat to remove water. Often the production of dry products requires several purification, evaporation, crystallization, and filtration steps.

There are many enzymes capable of hydrolyzing the glycosidic linkages in starch polymers (6,7). Amylases hydrolyze primarily α-1,4 linkages. Another group, debranching enzymes, specifically cleaves α-1,6 linkages at the branch points in starch polymers. The general action of each of these enzymes on amylose and amylopectin polymers is depicted in Figure 4-8.

α-Amylase is a very common *endoenzyme* that can be extracted from many sources including fungi, bacteria, mammals, and cereals. The mode of action, properties, and degradation products differ somewhat depending upon the source (8). For example, the temperature at which fungal α-amylase is inactivated is higher than that for cereal α-amylase. α-Amylase cannot hydrolyze the α-1,6 glycosidic

Endoenzyme—Enzyme that splits bonds anywhere along a polymer chain.

bonds that form the branch points in amylopectin, nor can it sever α-1,4 glycosidic bonds that are in close proximity to a branch point.

β-Amylase is an *exoenzyme* that specifically removes one maltose unit at a time from the nonreducing end of a starch molecule by hydrolyzing every other α-1,4 glycosidic linkage. The activity is arrested upon reaching the α-1,6 bonds of amylopectin (or before it reaches an α-1,6 bond if there are an insufficient number of glucose units for it to react), thus yielding a so-called *β-limit dextrin.*

Glucoamylase is an exoenzyme that removes glucose units from the nonreducing end of starch molecules. It has the ability to hydrolyze both α-1,4 and α-1,6 linkages. Hydrolysis of the α-1,4 linkages is considerably faster than that of the α-1,6 bonds. Eventually, however, if the glucoamylase activity lasts long enough, an entire starch molecule is completely converted into glucose, regardless of the extent of branching.

Debranching enzymes specifically hydrolyze α-1,6 glycosidic bonds. There are two enzymes that are commonly used to debranch starch, isoamylase and pullulanase, both commercially derived from bacteria (9).

When *cyclodextrin glycosyltransferase* (CDGTase) acts on starch or a starch derivative, the resultant product is a circular ring of glucose units linked by α-1,4 bonds (10). The enzyme severs the bonds within a starch substrate to create a linear chain of glucose units and rejoins the ends to form the ring structure. CDGTases from bacterial sources are used in the commercial preparation of *cyclodextrins,* which usually contain six (α-cyclodextrin), seven (β-cyclodextrin), or eight (γ-cyclodextrin) glucose units. Cyclodextrins are used commercially to form inclusion complexes with a wide variety of molecules. Many industrial applications involve the use of cyclodextrins to alter the apparent chemical and/or physical properties (e.g., solubility, volatility, or chemical stability) of the guest molecule in the inclusion complex.

Exoenzyme—Enzyme that splits only terminal bonds in a polymer chain.

β-Limit dextrin—Product resulting from the action of β-amylase on a branched starch polymer such as amylopectin.

Glucoamylase (α-1,4-glucan glucohydrolase)—Enzyme that removes the glucose units consecutively from the nonreducing ends of starch polymers by hydrolyzing both α-1,4 and α-1,6 linkages.

Debranching enzyme—Enzyme, such as isoamylase and pullulanase, that hydrolyzes the α-1,6 linkages of starch.

Cyclodextrin glycosyltransferase—Enzyme that cleaves starch and cyclizes the resulting glucose chain into a cyclodextrin ring.

Cyclodextrin—A circular molecule of α-1,4-linked glucose units.

Physical Modification

PREGELATINIZED STARCH

For the production of most pregelatinized starches (called pregels or instant starches), manufacturers have processing techniques that gelatinize the starch and then recover it as a dry powder prior to sale to the end-user. Pregelatinization methods discussed herein include drum drying, spray cooking, solvent-based processing, and extrusion. In most pregels marketed today, granular integrity has been lost. Granules are usually in fragments or a dissociated physical form. The instant starches marketed as cold-water swelling (CWS) types are unique in that they maintain granular integrity. These starches are prepared by either patented or proprietary processes. Pregels are generally used as thickeners in foods that receive minimal heat processing.

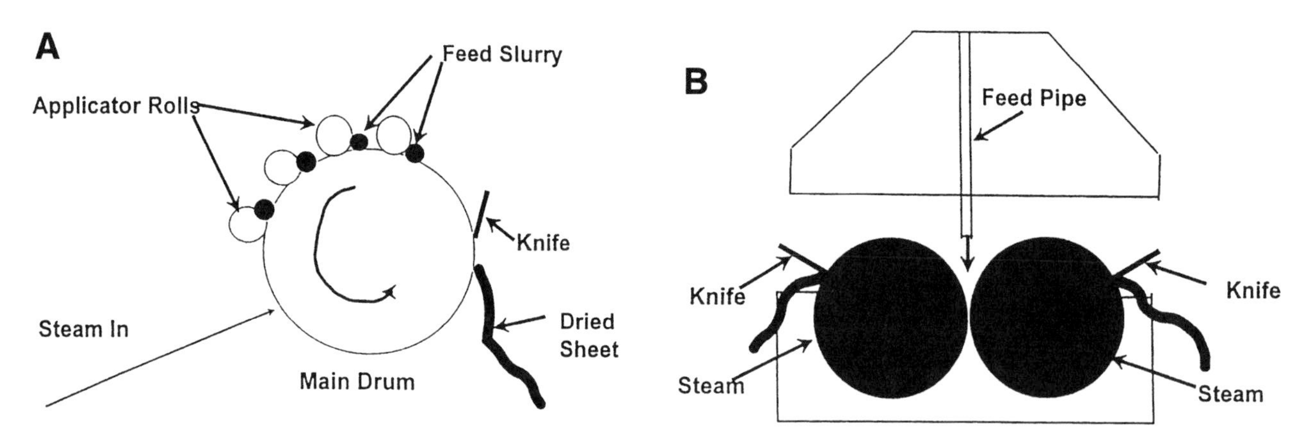

Fig. 4-9. A, Single drum dryer and **B,** double drum dryer.

Drum drying. Drum drying is perhaps the most common method of producing pregels. Although the process is conceptually straightforward, the operation of the drum dryer requires experience to produce high-quality and consistent products. In general, there are two common types of drum dryers, the single drum dryer and the double drum dryer. In the typical single drum process, a starch slurry at a concentration of about 30–40% solids is fed directly between the applicator roll and the heated drum (Fig. 4-9A). In the double drum process, the slurry is fed onto the heated surfaces at the gap position between the two drums (Fig. 4-9B). The cooked and dried starch film is removed from the drum surfaces by using a knife blade and ground to the desirable particle size. Generally, double drum drying has less mechanical shear and can produce thinner films. A dispersion of starch prepared from a double drum dryer tends to have a slightly higher viscosity than the corresponding product made on the single drum. It is also less grainy and pulpy, probably because of lower amounts of fragmented starch granules, the thickness of the sheet, and the degree of cooking. Other food ingredients such as gums, surfactants, sweeteners, salts, and flavors can be mixed into the starch slurry and processed during drying.

Viscosity profiles of pregels can vary depending upon the type of starch that is dried. The viscosity profile of a chemically modified, drum-dried starch usually differs from that of its pregelatinized native counterpart. Generally, the viscosity profile of a pregel shows "instant" viscosity (Fig. 4-10); i.e., the granule acts like a sponge to immediately imbibe water and produce viscosity. Depending upon the type of pregel, the viscosity may be reduced with longer cooking. "Precooked," and hence physically altered, pregels tend to retain some of the physicochemical properties of the starting base material. For example, if the pregelatinization step is done properly, most of the heat, acid, and shear stability of a crosslinked starch is retained. Perhaps the most significant changes that occur in starch upon pregelatinization (compared with its cook-up granular base starch) are in dispersibility and texture. These properties are usually a function of the particle size. Since drum-dried pregels are ground after drying, dispersibility and texture can be manipulated by controlling the grinding process and thus the resultant particle size. In general, fine-grind pregels are smoother in texture but more difficult to disperse. A coarse grind tends to generate pregels that have pulpy textures with better dispersibility.

Spray cooking. Pregels can be made by first cooking a starch slurry in a high-pressure, high-shear jet-cooker and then pumping it through a spray dryer. As the cooked starch passes through the nozzle of the dryer, it is atomized and dried. However, this method is not generally used commercially because of its high costs.

An alternative spray-cooking process omits the jet-cooking step and instead uses steam injection at the spray dryer nozzle to cook out the starch as it is being atomized. Starch produced by this method is uniformly cooked or gelatinized with a minimum amount of shear and heat damage. Because the pregelatinized granules remain essentially intact, upon rehydration they have the smooth texture and viscosity of a cook-up starch (11). These types of pregels are considered high-performance CWS products.

Solvent-based processing. A process for preparing pregelatinized starch by using aqueous alcohol was reported in 1971 (12) and put into commercial use during the mid-1980s (13). The process generally involves heating about 20% starch in an alcohol such as ethanol containing 20–30% water at approximately 160–175°C (320–350°F) for 2–5 min. The resulting product is generally referred to as a CWS granular starch. The granular integrity is maintained (unlike drum-dried or extruded starch products), but its birefringence is lost. The use of alkali in the process has been reported; however, it has not yet been adopted commercially.

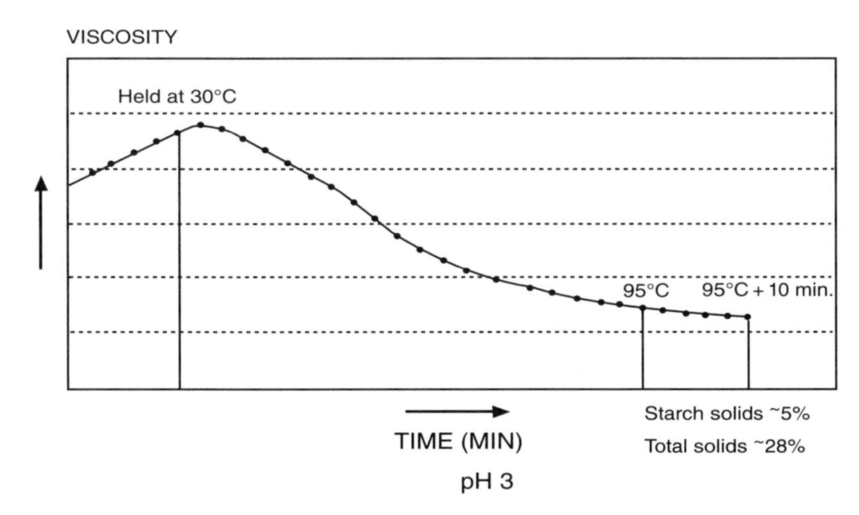

Fig. 4-10. Viscosity profile of a typical pregelatinized starch. (Redrawn from National Starch and Chemical Co., 1992, Instant Pure-Flo F, Tech. Serv. Bull. 12192-305)

Extrusion. Although extrusion processing can be used to produce pregels, it is not as widely used as the other processes. In very basic terms, the extruder can be regarded as a high-temperature reactor. Granular starch at various moisture levels is compressed into a dense, compact starch melt. The granules are destructured by the high pressure, heat, and mechanical shear encountered in the extruder. The starch extrudate exits through the die in an expanded form that results from the pressure differentials and flashing-off of internal moisture. It is then further dried and ground to meet specification criteria. The physical properties of the extruded starch are affected by the geometry, temperature, pressure, screw speed, and feed rate of the extruder and the moisture content of starch. Because of its high throughput and relatively low energy consumption, extrusion cooking is a relatively low-cost process for producing pregels, but in order to attain quality products with high viscosity and good texture, the technology needs further improvement.

HEAT-TREATED STARCH

A less common but nevertheless interesting physical modification process involves the heat treatment of starch. Under controlled conditions, granular starch can be heat treated and recovered in its granular form. Not only does the resultant starch maintain its cook-up properties, but it shows improved viscosity and stability when subsequently gelatinized and pasted.

There are two types of heat-treatment processes, heat-moisture and annealing, both of which cause a physical modification of starch without any gelatinization, damage to granular integrity, or loss of birefringence (14). The heat-moisture treatment involves heating starch at a temperature above its gelatinization point but with insufficient moisture to cause gelatinization. For example, in one patented heat-moisture process (15), tapioca starch is first adjusted to 15–35% moisture and then heated to 70–130°C (158–266°F) for 1–72 hr in a closed reactor. The change in the viscosity profile of tapioca starch after the heat-moisture treatment is shown in Figure 4-11. The resultant product has improved consistency and viscosity stability in foods with a pH of less than 4.5. Annealing of starch involves heating a slurry of granular starch at a temperature below its gelatinization point for prolonged periods of time. After annealing, the granular starch shows an enhanced viscosity profile.

It has recently been reported that when starch is heated under conditions of very low or no moisture, the resulting product provides functionality similar to that of chemically modified starch (16).

Although heat-treatment processes are not mainstream, they are unique in that chemical reagents are not required to impart a modifying effect. The resultant starch may be considered native and therefore labeled "food starch."

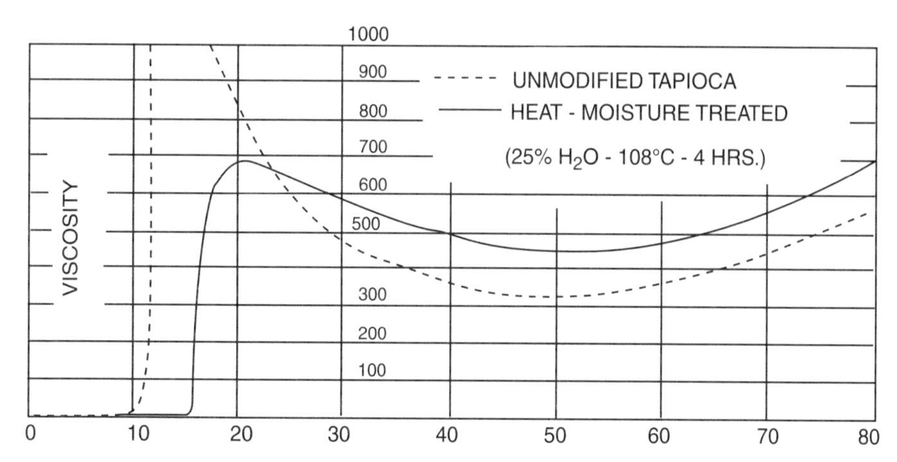

Fig. 4-11. Viscoamylograph profile of a heat-moisture treated tapioca starch. (From [15])

References

1. Whistler, R. L., BeMiller, J. N., and Paschall, E. F., Eds. 1984. *Starch: Chemistry and Technology,* 2nd ed. Academic Press, New York.
2. Rutenberg, M. W., and Solarek, D. 1984. Starch derivatives: Production and uses. Pages 311-388 in: *Starch: Chemistry and Technology,* 2nd ed. R. L. Whistler, J. N. BeMiller, and E. F. Paschall, Eds. Academic Press, New York.
3. Kirchoff, G. S. C. 1811. Page 27 in: Acad. Imp. Sci. St. Petersbourg Mem. 4.
4. Fleche, G. 1985. Chemical modification and degradation of starch. Pages 73-99 in: *Starch Conversion Technology*. G. M. A. Van Beynum and J. A. Roels, Eds. Marcel Dekker, New York.
5. Reeve, A. 1992. Starch hydrolysis: Processes and equipment. Pages 79-120 in: *Starch Hydrolysis Products.* F. W. Schenck and R. E. Hebeda, Eds. VCH Publishers, New York.
6. Kruger, J. E., and Lineback, D. R. 1987. Carbohydrate-degrading enzymes in cereals. Pages 117-139 in: *Enzymes and Their Role in Cereal Technology*. J. E. Kruger, D. Lineback, and C. E. Stauffer, Eds. American Association of Cereal Chemists, St. Paul, MN.
7. Teague, W. M., and Brumm, P. J. 1992. Commercial enzymes for starch hydrolysis products. Pages 45-78 in: *Starch Hydrolysis Products.* F. W. Schenck and R. E. Hebeda, Eds. VCH Publishers, New York.
8. Kulp, K. 1975. Carbohydrases. Pages 53-122 in: *Enzymes in Food Processing,* 2nd ed. G. Reed, Ed. Academic Press, New York.
9. Atwell, W. A., Hoseney, R. C., and Lineback, D. R. 1980. Debranching of wheat amylopectin. Cereal Chem. 57:12-16.
10. Hedges, A. R. 1992. Cyclodextrin: Production, properties, and application. Pages 319-333 in: *Starch Hydrolysis Products.* F. W. Schenck and R. E. Hebeda, Eds. VCH Publishers, New York.
11. Pitchon, E. P., O'Rourke, J. D., and Joseph, T. H. 1986. Apparatus for cooking or gelatinizing materials. U.S. patent 4,600,472.
12. Thurston, R. A., and McCoalga, R. E. 1971. Process for preparing granular cold-water-swelling starches and the starches resulting therefrom. U.S. patent 3,617,383.
13. Eastman, J. E., and Moore, O. 1984. Cold-water-soluble granular starch for gelled food compositions. U.S. patent 4,465,702.

14. Stute, R. 1992. Hydrothermal modification of starches: The difference between annealing and heat/moisture treatment. Starch/Staerke 44:205-214.
15. Smalligan, W. J., Kelly, V. J., and Enad, E. G. 1977. Preparation of a stabilized precooked baby food formulation thickened with modified tapioca starch. U.S. patent 4,013,799.
16. Chiu, C. W., Schiermeyer, E., Thomas, D. J., and Shah, M. B. 1998. Thermally inhibited starches and flours and process for their production. U.S. patent 5,725,676.

CHAPTER 5

Matching Starches to Applications

In This Chapter:

The versatility of starch, particularly modified starch, makes it well suited for a wide variety of foods (Table 5-1). Starch contributes to texture, viscosity, gel formation, adhesion, binding, moisture retention, film formation, and product homogeneity. Certain modified starches are also being increasingly used as fat substitutes in low- and no-fat products.

Commercial starches are obtained from cereals such as corn, wheat, and various rices and from tubers or roots such as potato and cassava (tapioca starch). Starches from different sources vary in taste and viscosifying properties. For example, native potato and tapioca starches have weak intermolecular bonding and gelatinize easily to produce high-viscosity pastes that thin rapidly with moderate shear. Potato starch produces clear, viscous, almost bland pastes, which are used in products such as extruded cereals and dry soup and cake mixes. Tapioca starch produces clear, cohesive pastes that gel slowly over time. Native corn, rice, and wheat starches form opaque, gelled pastes that have a slight cereal flavor. High-amylose corn starches produce opaque, strong gels commonly used in gum candies. Waxy maize starch produces a clear, cohesive paste (1).

Because the pastes and gels produced by native starches are often cohesive (gummy) or rubbery, the functional properties of these starches are improved by modification. Different types of modification produce starches that are better able to withstand the heat,

TABLE 5-1. Starch Usage by Application Sector[a,b]

Application	Binders	Viscosifiers	Film-Formers	Texturizers
Soups, sauces, and gravies	. . .	X, XS, PX, PXS	. . .	X, XS, PX, PXS
Bakery	PN	X, P, PX, PXS	D, M	P, X, PX, PXS, M
Dairy	N, A, M	X, XS, P, PX, PXS	. . .	X, XS, PXS, A, PX, O, PO, M
Confectionery	PO, O	. . .	O, PO, A	. . .
Snacks	N, P, PN, PO, D	. . .	. . .	. . .
Batters and coatings	X, PX, O	P, PX	D	O, PO, D, M
Meat products	N, X, XS, P	. . .	XS	XS

[a] Adapted from Davies, L., 1995, Starch—Composition, modifications, applications and nutritional value in foodstuffs, Food Technol. Europe, June/July, pp. 44-52.

[b] N = native, X = crosslinked, P = pregelatinized, S = substituted, O = oxidized, A = acid hydrolyzed, D = dextrin, and M = maltodextrin.

shear, and acids associated with various foods and food-processing conditions. Modification can have an effect on starch solubility, viscosity, freeze-thaw stability, paste clarity and sheen, gel formation, film formation, and cohesiveness.

There are numerous factors to consider in the choice of a starch for use in a particular food system. The desired properties of the food (e.g., texture, mouthfeel, and viscosity), the method of processing, and the distribution parameters, especially storage temperatures, must all be examined. An up-front strategy in which the various requirements of the food product are reviewed prior to selection of a starch saves time, frustration, and probably money in the product-development process.

General Considerations

SENSORY CONSIDERATIONS

The goal of the product-development process is to create safe, high-quality foods with optimal sensory properties. Sensory considerations such as texture, appearance, and flavor usually determine consumer acceptability. The desired "look" of a product, which is usually conceptualized prior to the development process, is not always easy to achieve. Ingredients, processing, and distribution all influence final product quality and ultimately dictate success or failure in the marketplace.

Since starch is an ingredient that can affect the texture, body, and appearance of a product, many factors need to be taken into account. Table 5-2 contains terms used in the vernacular of the starch industry that are associated with texture, body, and appearance of starch-containing foods. These terms, although not "scientific," have been used for years to describe the performance of starch, either as a paste or as part of a food product, during or after processing.

TABLE 5-2. Common Terms Used to Describe Starch-Based Food Products

Texture	Body	Appearance
Grainy	Heavy	Dull
Smooth	Thin	Shiny
Cohesive (gummy)	Long	Opaque
Gelled (soft or firm)	Short	Translucent

Texture. Starch can produce various textures, ranging from grainy to smooth and from cohesive (i.e., gummy or slimy) to gelled. Graininess is affected by the particle size of the starch granule itself as well as by the cooking characteristics. The particle size of drum-dried pregelatinized starches, for example, can be controlled by grinding so that various particle sizes are generated. Particle size not only affects texture, but dispersibility and dissolution of the starch as well. Powders of pregelatinized starch that have a fine granulation (i.e., small particle size and therefore a greater surface area) have a tendency to lump during hydration, especially if proper mixing is not achieved during the addition step. This problem is caused by quick hydration and gelling of the starch at the outer surface, thus preventing the interior starch from wetting. The centers of these lumps actually contain dry, pregelatinized starch powder. These characteristic gelled lumps are often referred to as "fish eyes" in the industry. Pregelatinized starches that are coarsely ground and/or agglomerated usually dissolve more easily because of their larger particle size (less surface area) and slower hydration rate. Large particles can sometimes generate a grainy, apple sauce-like texture, depending on how the starch has been modified, the manner in which the starch is pregelatinized, and/or its hydration properties. The amount of cooking that a starch undergoes may also affect texture. For example, cook-up starches that are not sufficiently gelatinized may result in products with a "pasty" or "grainy" mouthfeel. Finally, a pasty texture can also occur if too much starch is used.

Cohesive or gummy textures are most often related to the type of starch used. In particular, cooked pastes from native waxy starches often produce cohesive textures. Chemical modification, however, particularly crosslinking, generally changes a normally cohesive-type starch into one that generates a short, pudding-like texture. Gelling behavior is related primarily to the amylose content of the starch. In general, native starches with high levels of amylose have a greater tendency to form a gel after cooking.

Body. Related to texture is the flow property of a starch paste or final product once a force has been applied. Simply put, how does the product behave after processing, especially after being disturbed, e.g., stirred, spooned, mixed, blended, or poured? Some starches result in a paste and/or final product with a very short, heavy body or consistency (for example, a pudding or salve-like product) while others are considered long and thin, that is, less viscous and more syrup-like. Both the type of starch base and the type of modification influence the final texture and body of the product.

Appearance. Starch can significantly affect the appearance or clarity of the product. Certain foods are expected to be dull, while others may require a surface shine (gloss). Gelling starches, such as that from native dent corn, become opaque after cooking, and others, such as that from native waxy corn, are more translucent. Potato starch yields a clear paste. The chemical modification of starch can also affect paste clarity. For example, chemical substitution such as hydroxypropylation has a tendency to improve paste clarity.

Flavor. Another sensory consideration for the product developer is the flavor of the final product. Here, too, starch and starch-related products can have a dramatic impact. The starch base, the milling process, and the type of chemical modification of the starch can all affect the resultant flavor. Delicately flavored foods, such as cream sauces or lightly flavored puddings, can develop an off-flavor from their ingredients, including starch. Because starches are isolated from plant materials, off-odors associated with the botanical source from which they are milled may be present. Residual protein and/or lipid not extracted during the milling process can influence starch flavor. Cereal-based starches such as corn and wheat starch are sometimes considered to have off-notes described as "cardboard" or "cereal-like." Root and tuber starches such as tapioca and potato, on the other hand, are usually judged to be cleaner in flavor. The type of starch modification can also influence the flavor. For example, it is believed that acetylated starches can sometimes cause a recognizable off-note in delicately flavored foods. In general, when considering a starch for a specific application, one needs to keep in mind the desired flavor profile and understand the impact the chosen starch will have on that flavor.

pH CONSIDERATIONS

Another property of the food product that needs to be considered is pH. The pH of a product not only influences ingredient functionality, but is an important parameter in dictating food-preservation processes. For example, pH is a critical factor in determining the type of thermal process that is required for "commercial sterility" (as related to microbial growth and food safety). The thermal process requirements for high-acid foods (pH < 3.7), acid foods (pH 3.7–4.5), and low-acid foods (pH > 4.5) differ from one another and must meet strict guidelines (2).

Acidic conditions can have a dramatic impact on paste viscosity, texture, and stability. Native (unmodified) starches are typically unstable under acidic conditions. They are susceptible to viscosity breakdown during processing and may also result in the development of poor texture. Stability in an acidic environment is one of the primary reasons that starches are chemically modified. In general, the higher the degree of crosslinking, the more acid-tolerant the starch. Although neutral pH is less detrimental to starch viscosity, the thermal process that is usually required to preserve low-acid foods is quite severe. For this reason, modified starches are also required for low-acid products that require a more stringent thermal process.

FORMULA-RELATED CONSIDERATIONS

The food technologist also needs to be cognizant of other ingredients that could influence starch performance. Product composition should be kept in mind when determining the proper starch for a system. For example, the amount of moisture present to hydrate the starch, the presence of fats and oils, and the level of solids such as salts and sugars can all influence starch behavior. In general, ingredients that have a tendency to "complex" with starch and/or to compete directly with starch for water negatively impact starch performance, particularly during gelatinization and pasting. The effects of sugar content on starch gelatinization temperature and viscosity development are shown in Figure 5-1.

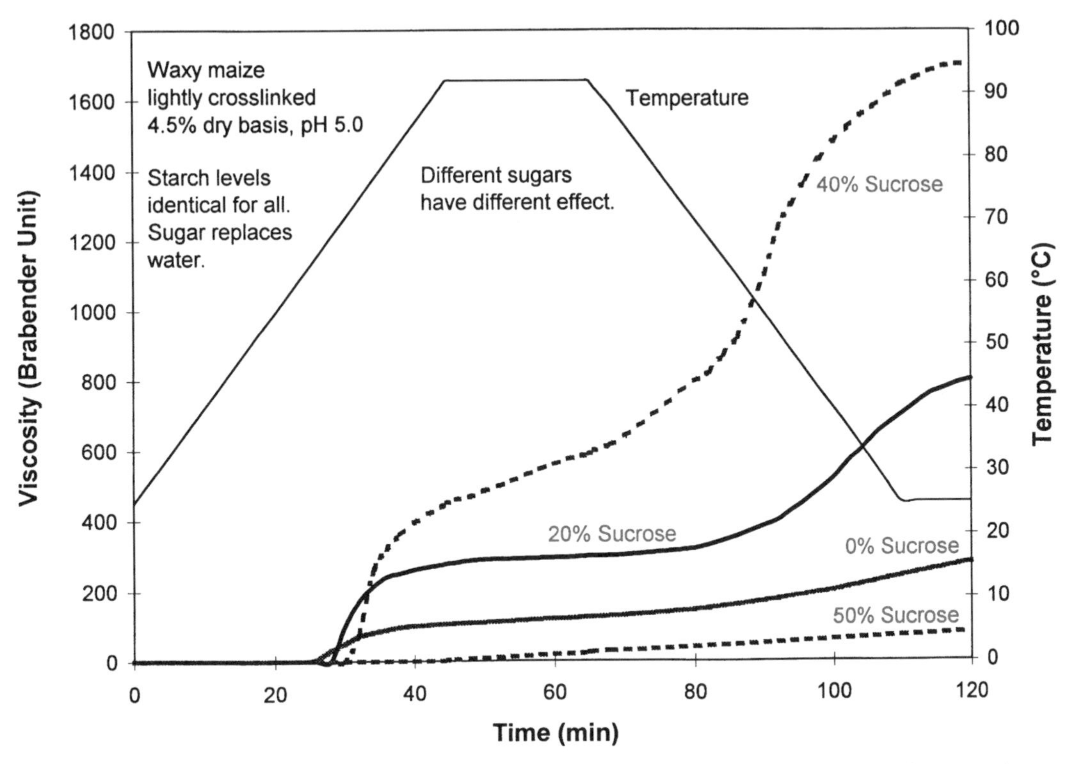

Fig. 5-1. Effects of sugar concentration on the gelatinization profile of waxy corn starch. Note that at the 50% sucrose level, almost no gelatinization takes place because little water is available. (Courtesy Cerestar USA)

PROCESSING CONSIDERATIONS

Although the selection of a starch that is compatible with a particular process is one of the most critical aspects of achieving proper starch performance (e.g., viscosity, gel formation, and water binding) and the desired product quality (e.g., texture and mouthfeel), it is perhaps the least understood. Starches are chemically modified not only for functionality in the product, but also for process tolerance.

Starches have been designed to perform in, or at least withstand, the usually hostile environment associated with various food-manufacturing processes. A key concern is the direct physical impact of the process. For example, processes that employ high shear, high heat, and/or high pressure are considered detrimental to starch performance. Thus, the type of processing equipment needs to be taken into consideration. High-shear pumps, mixers, homogenizers, colloid mills, and jet cookers can destroy starch integrity and significantly reduce paste viscosity. Therefore, in order to achieve proper functionality, it is critical to select a modified starch that can withstand such processes. In general, the higher the degree of modification, particularly crosslinking, the more resilient the starch and the better the chance that it will tolerate severe processing conditions.

TABLE 5-3. Conditions Created by Common Food-Processing Equipment[a]

Equipment	Processing Conditions
Pumps	Low to high shear
Steam-jacketed kettles	Low shear, long cooking times
Swept-surface cookers and coolers	Medium to high shear, medium cooking time
Steam-injection cookers	Medium to high shear, short cooking times
Jet cookers	High shear, high heat, short cooking times
Retorts	Low shear, high heat, high pressure, long cooking times
Extruder cookers	High shear, high heat, high pressure, short cooking times
Colloid mill	Very high shear
Homogenizers	Very high shear

[a] Adapted from (3).

Table 5-3 lists typical food-processing equipment and the conditions each creates for the food system (3).

DISTRIBUTION AND END-USE CONSIDERATIONS

The storage and distribution of the food product after processing must also be considered when selecting a starch. Probably the most critical factor is the temperature during distribution. Of particular concern are those products that are stored either refrigerated or frozen, especially if temperatures fluctuate during storage.

Without digressing into lengthy reviews on freezing curves, moisture migration, ice crystal growth, the effect of solutes, and types of cold-temperature storage equipment, it can be said that the primary requirement for starch in a refrigerated or frozen product is that it be chemically modified (see Chapter 4). Substituted starch contains chemical "blocking" groups linked to the starch polymers and is especially important for chilled or frozen applications. It is this blocking action that prevents retrogradation of amylose and amylopectin. Compared with native starch, substituted starch can better withstand moderate temperature cycling or protracted storage at refrigeration temperatures. Modified starches, particularly substituted starches, have what is commonly referred to as *freeze-thaw (F-T) stability*. The F-T stability of a food product is often tested by allowing the frozen product to reach ambient temperature and then refreezing it. This process is usually repeated through numerous cycles in order to ascertain changes in product quality through each cycle. Product quality usually refers to textural changes and/or a loss of water from the product, i.e., syneresis. If there is no appreciable loss in product quality, the system is said to be F-T stable.

Freeze-thaw (F-T) stability—Ability of a product to withstand cold temperature cycling and/or prolonged storage at reduced temperatures.

The end-use, i.e., how the product will be prepared by the consumer, must also be considered. For example, the choice of a starch will depend on whether the product will be microwaved or baked, reconstituted with hot water, or refrigerated or frozen after final preparation.

Starch-Selection Guides

In order to make starch selection easier for the food technologist, starch manufacturers have various types of booklets, pamphlets, and selection charts or tables available. The specific starch recommendations within these guidelines are usually functions of the food application (e.g., product texture and pH requirements), processing conditions (e.g., type of thermal process and amount of shear), and distribution requirements (e.g., storage temperature).

Given the multifunctional characteristics of starch, particularly certain types of chemically modified starch, one starch product may work adequately in a variety of foods. On the other hand, certain starches are specifically designed for a particular application. Just because a starch is "recommended" for a particular application, however, does not necessarily mean that it will always work appropriately. Unexpected variability, usually related to product formulation, processing, and/or storage conditions, can sometimes affect starch behavior. By using creative formulation and processing, it is sometimes possible to "force fit" a starch product into a particular food application for which it may have not been originally intended. Hence, selection information should be considered a best estimate of the proper starch and usage level and should not be considered absolute.

References

1. Whistler, R. L., and BeMiller, J. N. 1997. *Carbohydrate Chemistry for Food Scientists*. American Association of Cereal Chemists, St. Paul, MN.
2. Lund, D. 1975. Heat processing. Pages 31-92 in *Principles of Food Science II. Physical Principles of Food Preservation*. O. R. Fennema, Ed. Marcel Dekker, New York.
3. Smith, P. S. 1982. Starch derivatives and their use in foods. Pages 237-269 in *Food Carbohydrates*. D. R. Lineback and G. E. Inglett, Eds. AVI, Westport, CT.

CHAPTER 6

Confections and Dairy Products

Confections

In This Chapter:

When the role of starch in confections is discussed, the applications range from gum candies to hard candies. The former (e.g., gumdrops and jubes) are typically softer with a chewy consistency. The latter, as the name implies, are much harder and tend not to be chewable (e.g., lozenges). The gel-forming and texture-stabilizing abilities of starch are important in the manufacture of confections.

The type and level of starch used are crucial to quality, but other ingredients, such as sugar, corn syrup, fats and emulsifiers, and other solids, are equally important and greatly influence the texture and shelf life of the finished candy. For example, increasing the amount of sugar increases starch gelatinization temperature, and certain acids, such as citric, malic, and tartaric acid used to impart tartness, can disrupt hydrogen bonding and result in rapid starch breakdown, a weakened gel, and poor shelf life.

A concise review of the role of starches in the confection industry has been prepared by Zallie (1). Within this summary are details on manufacturing procedures and types of confectionery starches.

GUM CANDIES

Starch-containing gum candies are typically prepared by one of two methods: in an open kettle (Fig. 6-1) or in a high-temperature, high-shear continuous cooker, sometimes referred to as a "jet cooker" (Fig. 6-2). In the open-kettle method, the starch is first slurried and cooked separately at 91–96°C (195–205°F) and then combined with the other ingredients such as corn syrup and flavorings. The mixture is cooked until the desired solids level (approximately 70–80%) is achieved, and then the candy syrup is deposited into a mold and cooled under controlled temperature and humidity conditions. Candy operations that use a high-temperature, high-shear continuous cooker do not require that the starch and water be cooked separately. The entire candy mix is processed at temperatures of 141–168°C (285–335°F). It is obviously important to tailor the starch to the appropriate process.

Although many types of starch could be used depending on the application, two types of starch are most commonly used in the confectionery industry, high-amylose starch and *thin-boiling* (acid-thinned) *starch*. These starches can be used alone or in combination at various

Thin-boiling starch—Acid-hydrolyzed starch used to reduce the hot viscosity of a paste so that higher concentrations of starch can be dispersed without excessive thickening.

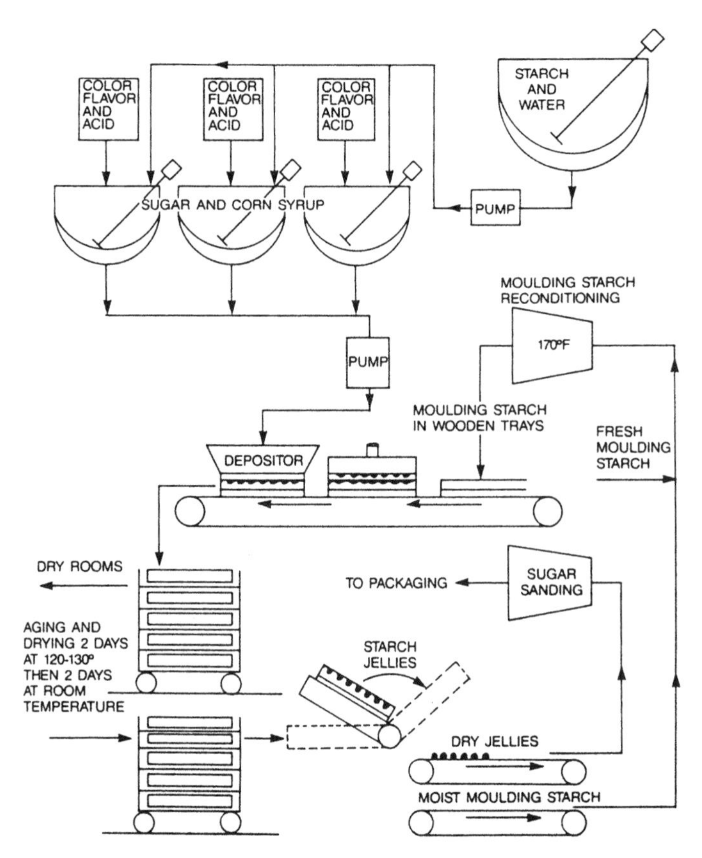

Fig. 6-1. Conventional kettle process for cooking of starch jellies. (Reprinted, with permission, from [1])

ratios depending on the texture desired. The total starch content present in starch-containing confections is usually about 9–15% by weight.

High-amylose starch, when properly cooked, has a high gel strength and sets up rather quickly (in about 24 hr compared with 48–72 hr for some other starches). This relatively rapid set provides the confectioner with improved productivity. Because high-amylose starch requires high temperatures (168°C [335°F]) in order to cook properly, a high-temperature cooker, such as a jet cooker, must be used. Although high-amylose starch provides high gel strength, it sometimes possesses "hot viscosities," which are too high to be properly handled. For example, if high-amylose starch is used at too high a level, tailing during deposition and rapid gelling, which can cause problems in the holding tank, can occur. For these reasons, high-amylose starches are often acid-thinned to reduce viscosity or used in combination with other thin-boiling starches.

The term "thin-boiling" refers to a starch with low hot viscosity, which allows concentrated sugar and starch solutions to be rapidly and efficiently cooked and handled (pumped and deposited into the

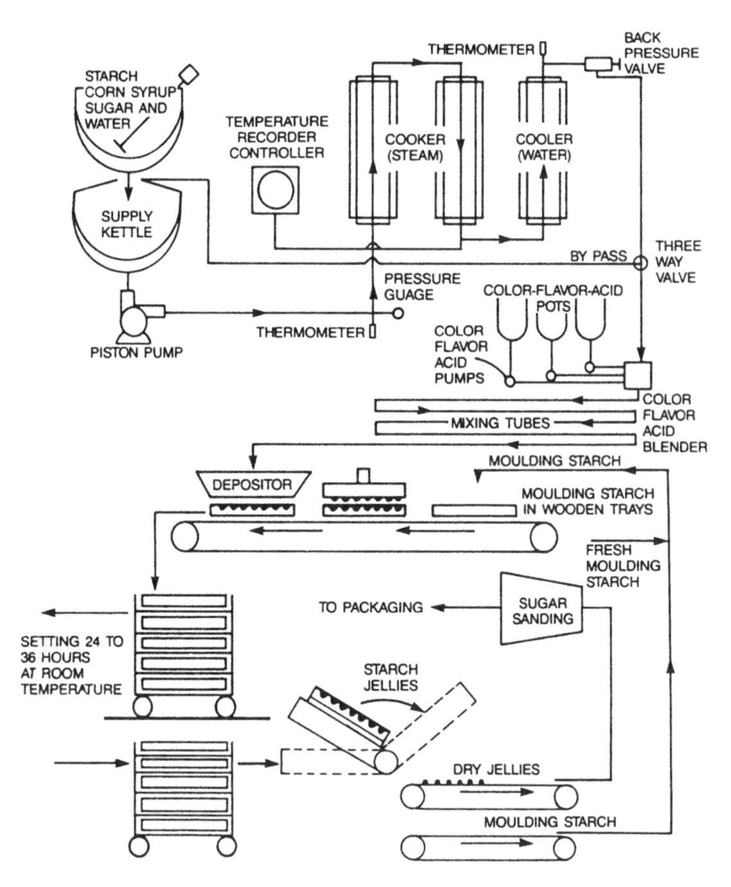

Fig. 6-2. Continuous process for jet cooking of starch jellies. (Reprinted, with permission, from [1])

molds). A thin-boiling starch is made by acid modification, which hydrolyzes glycosidic linkages within the granules, resulting in granules that contain starch polymers with reduced molecular weight. Reduction in molecular weight results in a starch that has low hot viscosity when initially cooked but then "sets up" to a gel upon cooling. This characteristic allows cooked candy syrups to be poured into molds and subsequently cooled into the desired shapes. The viscosity of thin-boiling starch (i.e., the degree of conversion) can be characterized by its *water fluidity (WF),* a term that describes the inverse of viscosity (see Chapter 4).

Pan coating. During pan coating, candies such as jelly beans are rotated in a revolving pan and coated with a concentrated sugar solution containing a film-forming agent. High-amylose, converted starches are often used in this coating for their moisture-absorbing and film-forming abilities. The coating plasticizes as it dries, preventing cracking and chipping of the coating and "sweating" of the jelly bean center.

Water fluidity (WF)—Scale used to determine the degree of starch conversion.

HARD CANDIES

Higher amounts of starch (typically 20–30% on a dry solids basis) are used in the manufacture of hard candies such as lozenges and cough drops. Because of these high usage levels, the starches must be lower in viscosity than those used in soft, jelly-type candies. Very low viscosity, thin-boiling starches and dextrins can be used to provide a pliable toughness to the candy piece rather than the cuttable gel structure required in soft candies.

NOUGATS, CARAMELS, AND TOFFEES

Nougats are combinations of sugars, syrups, and whipping agents. There are basically two types of nougat textures, short and chewy, controlled by using different amounts of noncrystalline sugars and the addition of a specialty starch. Caramels are similar to nougats except that they contain milk solids. The texture ranges from chewy to short. Toffees are high-cooked caramels with textures ranging from chewy to brittle and crunchy.

Specialty starches are available for all these candy applications and differ depending on the property desired. When chewiness is desired, an unmodified waxy starch (usage level approximately 2%) or a native tapioca starch are among the types that can be used. When stiffness, body, and textural stability are desired, a modified specialty starch (approximately 3%) derived from waxy starch can be used. Higher usage levels (4–6%) of specialty starches provide firmness and rigidity and the rapid set needed for the continuous extrusion of these products. These starches cook at the low temperatures at which these products are produced.

STARCH MOLDS

Preformed starch molds are used in the confectionery industry to give the candy its desired shape and to absorb moisture from the cast candy as it dries, cools, and sets. The molds are formed from starch that has been treated with a small amount of mineral oil to minimize dust formation and enable the starch to accept and hold impressions. The moisture level of the molding starch is critical. At levels below 6%, the starch tends to remove water too fast, causing crusting or hardening. At levels above 9%, drying time is increased and the candy exteriors become crusty. The moisture content of the molding starch is the often overlooked cause of low-quality candy.

MARSHMALLOWS

Marshmallows are a confection generally made by adding an aerating agent (such as gelatin) to sucrose or glucose syrups and then pouring the mixture into starch molds. Marshmallows can also be made by extrusion. In extrusion-type formulations, starch is often added as an ingredient in the formula. The starch aids in the cutting and forming operation and helps the final product retain its shape. The texture of a marshmallow depends on the final moisture content (12–18%), the aerating agent, and the last processing step (starch molding or extrusion). One of the defects that can occur when starch is used is a final product with a long, stringy texture. This problem can be avoided by using a modified starch with sufficient crosslinking.

Dairy Products

Although pudding is perhaps the classic example of a starch-containing dairy food, starch is also used in other dairy applications such as frozen desserts, yogurt, sour creams, and cheese sauces. Modified starches are typically used in dairy products at about 1–7% (by weight) of the total formula and function mainly as texturizers, stabilizers, and/or viscosifying agents.

Of primary concern from a starch-selection point of view are the processing and distribution requirements that are unique to the dairy industry. Three features often required of starches used in dairy products are that they be 1) heat tolerant, 2) shear tolerant, and 3) freeze-thaw stable. Given these requirements, it is generally the case that the best-suited modified starch is one that is both crosslinked and substituted. From within this category of starches, however, it is still necessary to select a starch that best fits the system. Dairy foods can undergo various types of heat treatment as well as high-shear homogenization. Different modified starches are often necessary for high-temperature short-time (HTST), low-temperature long-time (LTLT), and ultra-high-temperature (UHT) pasteurization (2). Although starch gelatinization and pasting is time and temperature related, a starch with a higher degree of crosslinking may be required for a process with a higher temperature, i.e., UHT versus LTLT. It is also critical that the level of crosslinking be sufficient so that the starch is able to withstand the high shear of the homogenization process. A final concern in selecting the proper starch is the distribution and handling of the final product. The fact that many dairy foods are held in cold-temperature storage requires that the starch be substituted, i.e., freeze-thaw stable.

Dry mix applications, such as instant puddings or instant cheese sauces that are prepared by the addition of milk or water, typically see little if any heat treatment and are therefore usually formulated with pregelatinized starches with a low to moderate level of crosslinking. Since these products are sometimes chilled after preparation, freeze-thaw stability is still a requirement.

Acid tolerance is required of starches that are formulated for use in fermented or acidified dairy foods. The most common example of a fermented dairy product is probably yogurt. A mixed culture of *Lactobacillus bulgaricus* and *Streptococcus thermophilus* is commonly used to ferment milk into yogurt. The milk sugar lactose is the primary carbohydrate source, and the key by-product of this microbial fermentation, lactic acid, causes a drop in pH, which in turn results in aggregation or "curdling" of the milk protein *casein*. It is this aggre-

Casein—A milk protein that is a complex of protein and salt ions, principally calcium and phosphorus, present in the form of large, colloidal particles called micelles.

gation of the casein micelles that produces the typical yogurt-like coagulum. The final pH of yogurt is approximately 4.0. Starch provides texture and stability to yogurt and helps prevent or limit "wheying-off" (syneresis), a quality defect related to excess serum separation from the coagulum. In low-fat yogurts containing fruits, starch helps build texture to suspend the fruit or improves the appearance of the fruit-on-the-bottom yogurts by keeping the yogurt out of the fruit portion.

In frozen dairy desserts, starches provide viscosity and a smooth, creamy texture. In low-fat products, they can be used to replace fat or oil.

Because dairy products usually possess a delicate flavor profile, bland-tasting starches are usually preferred. Tapioca-based starch products, for example, are considered clean-flavored and are particularly useful in these applications.

Troubleshooting

This troubleshooting section lists some common problems, causes related to starches, and suggestions for changes to consider in the formulation and processing of confection and dairy products. However, a problem is often the result of a combination of factors that have to be considered: the amount and type of starch used and the kind of modification it has undergone, processing conditions (e.g., heat, shear, moisture, and time), effects of other ingredients such as fat and sugar, and factors such as pH and storage conditions. The solution may involve any one or more of these factors. This guide may serve as a starting point in looking at specific products.

CONFECTIONS		
Symptom	**Possible Causes**	**Changes to Consider**
GUM CANDIES		
Candy not set properly	Insufficient starch gelatinization	Increase cooking time and/or temperature. Adjust water content. Check solids level. Check drying room conditions.
	Not enough starch	Increase starch content.
Hard outer shell	Molding starch moisture too low	Adjust molding starch moisture content.
	Drying room temperature too high; relative humidity too low	Adjust drying room conditions.
Gel forms too rapidly	Starch thickens too early during processing	Adjust ratio of high-amylose starch to acid-thinned starch or modified high-amylose starch.
Dry, firm, dull appearance	Incomplete gelatinization of starch	Increase cooking time and/or temperature.
	Too much starch	Decrease starch content.

Symptom	Possible Causes	Changes to Consider
Elastic, limp, shiny appearance	Overcooked starch; overprocessed candy	Decrease cooking or hold time and/or temperature.
	Not enough starch	Increase starch content.
CHEWY CANDIES		
Texture too short	Improper sugar crystal formation due to available water	Increase cooking time and/or temperature. Select a starch with less chemical crosslinking.
Candy too soft, does not hold shape	No structural matrix	Select a modified starch with sufficient chemical crosslinking.
HARD CANDIES		
Too brittle	Structure lacks adequate water	Select a modified starch with appropriate chemical crosslinking; i.e., with more water-holding capacity. Adjust water content.
Chewy	Structure contains too much water	Select a modified starch with appropriate chemical crosslinking; i.e., with less water-holding capacity. Adjust water content.
MARSHMALLOWS		
Hard outer shell	Moisture migration	Optimize starch level and gelatinization. Use a modified starch with better water-holding capacity.
Long, stringy texture	Overcooked starch	Use modified starch at proper level and gelatinization. Decrease cooking time and temperature. Decrease shear during processing.
Poor shape retention; poor cuttability	Poor structural matrix	Increase starch content. Decrease cooking time and/or temperature. Use modified starch. Decrease shear during processing.

DAIRY PRODUCTS

Symptom	Possible Causes	Changes to Consider
GENERAL		
Syneresis; weeping	Poor freeze-thaw stability; colloidal system breakdown; increased water mobility	Use substituted (stabilized) starch. Decrease shear during processing. Increase starch level. Increase cooking time and/or temperature.
Runny texture	Low solids level	Increase starch level. Monitor cooking time and/or temperature. Check starch selection. Decrease shear during processing.
Graininess	Starch not cooked	Check starch selection. Adjust processing temperature or time. Adjust water content.
YOGURT		
Too runny	Low solids level; acid hydrolysis of starch present	Select a modified starch with appropriate chemical crosslinking. Increase starch level.

Product settling	Poor suspension of solids	Select a modified starch with appropriate chemical crosslinking. Increase starch level.
Syneresis; weeping	Poor freeze-thaw stability; colloidal system breakdown; increased water mobility	Use substituted starch. Increase starch level. Decrease shear during processing.

References

1. Zallie, J. The Role and Function of Specialty Starches in the Confection Industry. Tech. Bull. National Starch and Chemical Company, Bridgewater, NJ.
2. Zallie, J. Specialty Starches in the Dairy Industry. Tech. Bull. National Starch and Chemical Company, Bridgewater, NJ.

CHAPTER 7

Grain-Based Products

In This Chapter:

Bakery Products

The role of starch in baked goods is usually associated with the native starch found in wheat flour. For example, most commercial wheat varieties range from about 8 to 16% protein and about 71 to 79% carbohydrate, which is primarily starch. Whether a hard (high-protein) wheat flour is being used for making bread and pasta or a soft (low-protein) wheat flour for cakes, cookies, and crackers, the key point is that the characteristics of the final product usually depend upon the unique synergy between the starch, protein, and other ingredients present in the dough or bakery mix. Starch behavior in these products is a function of the type of flour used, the product formulation (i.e., the other ingredients such as salts, sugars, emulsifiers, and shortening), processing conditions, and final baking requirements.

Adding modified starches to baked goods provides benefits that might not come from the use of flour alone. Given the limited amount of moisture in baked goods, particularly after baking, the primary benefits of adding modified starches are moisture retention and improvement of the texture. They also improve cell structure, increase volume and machinability, enhance shelf life, and keep particles from settling. For example, a pregelatinized starch with the ability to swell in unheated water is used in products such as thin muffin batters containing particles (e.g., blueberries) that would otherwise settle to the bottom before the thickening effect of the wheat starch is achieved during baking.

Modified starches, particularly lightly crosslinked and substituted pregelatinized starches, are perhaps the most functional in baked goods. A pregelatinized starch helps bind what relatively little moisture is present, thus providing improved tenderness in the final baked product, and contributes to the development of a fine, uniform cell structure.

A problem common to almost all baked goods is staling. In fact, staling begins as soon as baking is complete and cooling begins. Substituted pregelatinized starches help baked goods retain moisture after baking, thus extending shelf life.

CAKES, MUFFINS, AND BROWNIES

Cakes, muffins, and brownie-type product batters are emulsions of starches, proteins, sugars, water, and fats. Although the formulations may vary, the basic ingredients are flour, sugar, eggs, shortening, a leavening agent, salt, and a liquid such as water or milk. There are several types of cake formulations ranging from a high-ratio cake (i.e., one having a higher proportion of sugar than flour) to a dense brownie to a foam-type angel food cake. Starches in these formulas serve several functions. Upon heating, the granular starches present in the formula first swell, absorbing water. This increases the viscosity and therefore the stability of the batter emulsion. In cake-type products, there is enough water present to gelatinize the starches, although some ungelatinized granules may remain. Gelatinized starches form a matrix in which air bubbles are entrapped. This matrix is a major structural component in the cake cell structure. The timing of the gelatinization of the starches is important to ensure proper final cake structure and crumb development. If the starch gelatinization takes place too early, a low cake volume results. Also, the edges can set too rapidly, resulting in a cake that rises too high at the center. A low cake volume can also result if starch gelatinization is inhibited or occurs too late, because the leavening reactions take place without a cell structure to support them. Thus, the cake may initially rise but then collapses and shrinks. Gelatinization timing can be controlled by changing the level of sucrose or other sugars, such as dextrose, in the formula or by adding unmodified or modified starches, depending on the formulation.

The basic ingredients of cake, muffin, and brownie batters are flour, sugar eggs, shortening, a leavening agent, salt, and a liquid such as water or milk.

Brownies can be either cake- or fudge-type products. In cake-type brownies, the starches gelatinize and provide a cell structure and crumb softness similar to that of a typical sponge cake. The sugar content is much higher in the fudge-type brownies, and thus the extent of gelatinization is not as great as in a cake formula. The final product collapses after baking, is much more dense, and has a moister crumb.

Starches are added to batter formulations at a level of about 1–5% and can affect several aspects of the final product. Starches are often added as tenderizers in order to soften the crumb and as thickeners to increase the viscosity of the batter, which improves the fineness of the air cell size and increases final product volume and uniformity. Pregelatinized waxy starches, for example, are often used in dry cake mix applications because of their ability to disperse during mixing of the batter. Crosslinked starches help provide shear resistance, which improves tolerance to overmixing by the consumer. The addition of starches also aids in moisture retention in the final product, a key product attribute in extra-moist or pudding-type cake mixes.

Modified starches can be used as fat mimetics or fat replacers and are often used in reduced-fat cake applications (see Chapter 9).

COOKIES

Cookie formulas can vary dramatically depending on the desired final product. Typical cookie ingredients include flour, water, eggs, cocoa or chocolate, sugars, fats, and leavening agents. The starch present in cookies comes from the flour fraction, unless starches are added for specific functionalities. The cohesiveness of a cookie dough and the framework of the final product are derived from the starch. Abboud and Hoseney (1) showed that the native starch in a typical cookie formula remains almost entirely ungelatinized after baking and is present as intact granules. The starches do not gelatinize because of the presence of sugars, which compete with the starches for the water needed for the gelatinization process. Damaged starch granules can absorb water from the other ingredients present, affecting the spread of the final product. A decrease in cookie spread occurs as the degree of damaged starch increases.

Typical cookie ingredients include flour, water, eggs, cocoa or chocolate, sugars, fats, and leavening agents.

Starches are often added to formulations as inexpensive tenderizers to create soft, chewy cookies. These added starches are often waxy starches that have been modified to increase their water-holding properties. These starches help to form a tender structure and reduce the stickiness of the dough, which can aid in the molding and cutting processes. The starches can also help to reduce the oiliness of the final product by replacing some of the shortening used in the formula.

FILLINGS AND GLAZES, ICINGS, AND FROSTINGS

Many bakery products contain fillings or are coated with some type of topical application such as a thin glaze or rich frosting. In many of these applications, starches play an important role. In general, starches are able to increase the viscosity, which contributes to the creamy and smooth mouthfeel. They can also increase the freeze-thaw stability of the product by retaining the moisture and inhibiting moisture migration (syneresis). Specific types of starches are more suited to certain applications than others.

Fillings. Fillings can be separated into two major types: fruit fillings and cream fillings. Fruit fillings generally are translucent, with color coming from the fruit source. They can be produced with or without heat and are commonly in the acidic pH range of 3–4.5. Therefore, the starches used in these applications should be able to form a clear gel and withstand acid pH levels and possibly be tolerant to heat. Waxy starches meet these criteria and are often used.

For cooked or canned fruit fillings, starches are commonly used at 3–5% to provide viscosity; smooth, short textures; clear (or translucent) gels; good heat and acid stability; and stability under cold storage conditions. These starches are usually crosslinked to increase resistance to shear and stability under acidic or high-temperature processing conditions.

For instant fruit filling applications, pregelatinized starches are often used. The pregelatinized starch offers ease of dispersibility during mixing and provides instant viscosity without cooking. For fillings that will be frozen, acetylated or hydroxypropylated starches can be used because of their improved water-holding capabilities and resulting freeze-thaw stability.

Fruit fillings in filled or topped cookie applications are often very dense, with soluble solids of 70–80%, and have a low pH. In this application, starches are commonly used at a level of 5–8% to help provide a pulpy texture and to stabilize viscosity during baking.

For cream fillings, it is important that the starch used does not mask or alter the delicate flavor of the filling. Tapioca starches are commonly used at a level of about 3–7% in these applications because of their good, clean flavor profile. They also provide smooth, short textures and a creamy mouthfeel. Pregelatinized versions are also used for instant cream filling formulations to provide stability under cold conditions and adequate viscosity to control the migration of moisture from the filling to the bakery product.

Glazes, icings, and frostings. There are no clear definitions for the terms "glazes," "icings," and "frostings" in the bakery industry. For this discussion, glazes and icings are considered mixtures of sugars and other ingredients such as milk or water that are heated and applied to a product to form a very thin layer. Frostings are considered thick coating applications made of sugars, shortenings, and other flavoring or texturizing ingredients. In all these applications, starches and starch conversion products (e.g., maltodextrins or starch syrup solids) can help in forming a thin, clear, nontacky film on the product. They also help to soften the surface texture so that a hard shell does not form. Starches or starch products can also act as stabilizers by helping to reduce the moisture migration from the icing into the product. Starch glazes are also used to adhere particles (salt and seasonings) to the surfaces of products such as crackers. In canned and dry mix frostings, starches are commonly used at a level of 1–3% to provide a smooth, creamy texture and enhance moisture retention.

Extruded Products

There are many types of extruded foods available in the marketplace today. Familiar products include pasta, breakfast cereals, salty snacks, and pet foods, and starches play a very important role in each of these extruded foods. Starches gelatinize with the heat, shear, and water present during the extrusion processes. Manipulating starch gelatinization by changing the formulation (e.g., sugars, salts, and water) or the processing parameters (e.g., temperature and shear) influences the product characteristics of crispness, bite, expansion or puff, and texture and product-specific attributes such as the bowl life of cereals.

Common extruded products include pasta, breakfast cereals, snacks, and pet foods.

In developing a formulation for a particular extruded product, the process and the choice of a particular starch are very important because they are directly related. First, we will consider the starch. Both the amylose and amylopectin components of starch contribute to the attributes of the final products. Although formulations vary, generalizations can still be made regarding the impact that these starch components have on the final product. The amylose portion is typically responsible for the strength of the mix and crispness of the final products. Increasing the amylose fraction in the formulation makes the molecules more resistant to shear degradation during extrusion and can help to improve cutting and shape retention during drying or during the final processing steps such as baking or frying. High-amylose starches form a tight polymer network after gelatinization, which can help to improve the bowl life of a cereal by decreasing the ability of the milk to penetrate the matrix.

When starch gelatinizes, the branched amylopectin forms a web-like network within the food matrix. The network increases the product viscosity and forms a structural framework for the expansion process. The setting of the framework aids in providing a crisp final product. However, because of the polymer branching, amylopectin is susceptible to shear degradation during extrusion. The resulting dextrins and short-chain polymers can cause an increase in stickiness and difficulty during cutting operations after extrusion.

Starch selection is critical for controlling the amylose-amylopectin content. Since granule shape and amylose-amylopectin ratio differ among starches, the expansion property of each starch is unique. For example, waxy maize starch has a higher concentration of amylopectin than common corn starch and therefore can contribute to an increase in the expansion of an extruded product. Potato starch has the largest granules of all commercial starch types and also the greatest swelling capacity. However, an extruded dough made with a starch with such a high swelling capacity is sometimes difficult to roll into thin sheets of uniform thickness.

Native starches are generally not resistant to the high temperatures and high-shear processing conditions involved in the production of extruded snack foods, and breakdown of the starch granules results in a sticky dough with reduced water-holding capacity. Crosslinked starches are more resistant to shear degradation while still providing a network for dough expansion and are often used in formulations to help solve this problem. Therefore, light to moderately crosslinked starches are used to provide better stability. Excessive crosslinking lowers the swelling capacity of the starch, resulting in a final product that has a tendency to be less puffy, not uniform, and poor textured.

The second factor that is important to consider in the production of an extruded food is the type of processing used. There are two main types of extrusion processes, the indirect expansion process and the direct expansion process. Indirectly expanded foods are produced commonly by single-screw extruders. The extrudates are low-moisture, shelf-stable, intermediate products in the form of pellets often referred to as "half-product," implying that the product is half way through its production process. The half-product is commonly finished by either baking or frying prior to final packaging and consumption. The moisture content of the dough can range from 15 to 30%, and the temperature (75–120°C [167–248°F]) and shear of the process are relatively mild. As discussed previously, starch gelatinization is the key to product expansion. Because of the limited moisture and heat, pregelatinized starches are often used in indirectly expanded product formulations. However, because of the loss of starch granule integrity, pregelatinized starches may result in sticky, gummy doughs that do not extrude well and do not retain their shape after cutting or forming. Therefore, pregelatinized starches that maintain their granular shape (cold water swelling starches) are often used to improve the handling and forming properties of the dough used to make the half-product.

Directly expanded products are commonly made with the twin-screw extruder. Because of the greater versatility of the twin-screw design, several different products can be produced with one machine. In general, the moisture content of the dough is about 15–20%, and the temperatures range from 160 to 180°C (320 to 356°F) in the cooking zone of the extruder. Compared with the single-screw extruder, the typical twin-screw process subjects the mix to much higher shear, temperature, and pressure conditions. The mix in the twin-screw extruder also expands readily as a result of the flash-off of the steam and the sudden pressure drop at the exit of the extruder. Hence, the product directly expands.

The starch can undergo shear degradation during the twin-screw process. "Cook-up" rather than pregelatinized starches are preferred for directly expanded products. Crosslinked starches are also often used to help combat the detrimental effects that high shear, pressure, and temperature can have on the mix.

Batters and Breadings

Starch is present in batters and breadings because of both its native presence in flours and its use as an individual ingredient to enhance the coating performance. There is currently a wide variety of commercial starches recommended specifically for batters and breadings. Performance of these products is dependent upon the food material to be coated, the type of coating system itself, and the cooking process. Food products coated with batters and/or breadings include seafood, poultry and meat, and plant materials such as onions, mushrooms, and potatoes. French fries, for example, have become extremely popular, and coatings that range from clear, innocuous ones that improve holding times under heat lamps to darker and spicier ones that result in dramatic flavor and textural changes are used.

In many batter and breading applications, the food product is fried. Frying is a unique cooking process that is essentially an exchange of the water present in the food for the frying oil. The use of modified starch in batters and breadings for fried applications is quite common and can help to improve texture and appearance while reducing oil pick-up.

A batter can range in viscosity from a rather thin, pourable liquid to a thick mixture or suspension. Although similar ingredients are used, a breading is generally thought of as a mixture of ingredients that is applied as a dry blend to a moist food product prior to cooking. An excellent review of batters and breading was published by Kulp and Loewe (2). Types of starches used in batter and breading systems range from high-amylose starches to converted starches to pregelatinized starches (Table 7.1).

A typical batter formula is shown in Table 7.2. The addition of starches to formulations is shown in the table as a level of 0–5%. However, in some formulations, this level can be as high as 15%. Starches play an important role in the film-forming ability of these formulations and in reducing oil pick-up during frying. They also provide viscosity, good adhesion and binding properties, and uniform coating thickness and aid in maintaining integrity during freezing, thawing, frying, baking, and microwaving. Sensory characteristics including tenderness, crispness, and thickness can all be modified through starch choice. In the batter systems, the water present in the formula allows the starches to gelatinize with heat. The gelatinization is responsible for the increase in viscosity. The starch should be evenly distributed before gelatinization occurs so that the final coating is of uniform thickness.

Specialty starches, including pregelatinized, chemically or physically modified, and crosslinked starches, are available that modify specific key product attributes. Pregelatinized starches help to immediately build viscosity in batter systems, which aids in the application of the coating. Starches can also be modified so as to improve water retention, decrease oil pick-up during frying, and aid in the suspension of solids such as spices or bread crumbs. High-amylose starches aid in film-forming properties, which help to retain moisture in the food system and provide a continuous film matrix during further heat processing. They are also important for reducing oil pick-up in fried foods and increasing coating crispness. Converted starches, such as dextrins and oxidized products, provide uniform coating and are considered to have good adhesion properties.

TABLE 7-1. Recommended Starch-Based Products[a] for Batters and Breadings

Type	Functional Attributes
High-amylose products	
Starch	Reduction in oil pick-up, crispness,
Flour	brittleness, film formation
Converted starches	
Dextrins	Uniform coating, adhesion,
Oxidized starches	crispiness, film formation
Pregelatinized starches	
Crosslinked	Viscosity control
Stabilized	

[a] Includes native and modified products.

TABLE 7-2. Typical Formulations of Batter Systems[a]

Ingredient	Addition Range (%)
Critical	
Wheat flour	30–50
Corn flour	30–50
Sodium bicarbonate	Up to 3
Acid phosphate	Adjust, based on neutralizing value
Optional	
Flours from rice, soy, barley	0–5
Shortening, oil	0–10
Dairy powders	0–3
Starches	0–5
Gums, emulsifiers, colors	Less than 1
Salt	Up to 5
Sugars, dextrins	0–3
Flavorings, seasonings, breadings	Open

[a] Reprinted from (2).

Troubleshooting

This troubleshooting section lists some common problems, causes related to starches, and suggestions for changes to consider in the formulation and processing of grain-based products. However, a problem is often the result of a combination of factors that have to be considered: the amount and type of starch used and the kind of modification it has undergone, processing conditions (e.g., heat, shear, moisture, and time), effects of other ingredients such as fat and sugar, and factors such as pH and storage conditions. The solution may involve any one or more of these factors. This guide may serve as a starting point in looking at specific products.

Batter or dough viscosity can also be influenced by the type of flour used and the flour- or starch-to-water ratio. Be sure that bread flour is used in yeast-raised products and cake flour in cakes and cookies.

BAKERY PRODUCTS

Symptom	Possible Causes	Changes to Consider
GENERAL		
Tough crumb structure	Overdevelopment of gluten structure	Add starch to disrupt gluten structure.
Runny batter; sticky bread or bagel doughs	Too high water level; low flour or starch level	Decrease amount of water added. Add starch or flour to increase solids level. Decrease mixing times.
Poor, nonuniform cell structure; low volume of final product	Gas not retained in baked system	Add waxy starch. Add flour.
Product lacks desired moistness	Moisture migration out of product	Increase water content. Select a substituted or waxy starch.
CAKES		
Collapsed structure	Insufficient starch gelatinization	Decrease sweetener level. Increase water level.
Low cake volume	Starch gelatinized too early	Increase sucrose content. Select a starch with more crosslinking.
	Insufficient starch gelatinization	Decrease sucrose. Increase oven temperature; decrease baking time. Increase water level.
Overrisen center (dome-shaped cake)	Starch gelatinized too early	Increase sucrose content. Select a starch with more crosslinking.
BROWNIES AND DRY MIXES		
Cake-like brownie (when fudge-like desired)	Starches fully gelatinized during baking	Increase sucrose content. Increase starch level. Decrease flour level.
Fudge-like brownie (when cake-like desired)	Starch gelatinization inhibited	Decrease sucrose content. Decrease starch content. Increase flour level.

Symptom	Possible Causes	Changes to Consider
Lumpy batter; poor dispersion of ingredients	Low wettability of dry ingredients	Preblend pregelatinized starch with flour. Select a pregelatinized starch with coarser particles.
COOKIES		
Crumbly, dry	Low cohesiveness in dough	Add pregelatinized starch and increase water level.
Increased spread	Low absorption of water by starch	Increase starch level.
Hard (soft, chewy cookie desired)	Low level of water absorption; low tenderization	Add pregelatinized starch.
Poor cuttability	Sticky dough	Decrease water content. Check starch selection.
Fat deposit (grease mark) on napkin from cookie	Dough too oily	Replace a portion of the shortening with starch.
FILLINGS AND GLAZES, FROSTINGS, AND ICINGS		
Low viscosity	Too low level of thickening agent	Add starch.
Grainy texture	Poor mixture, emulsion	Add pregelatinized starch.
Syneresis	Moisture migration	Add pregelatinized starch.
Cracking; formation of a hard shell	Dehydration	Add modified starch with better water-holding capacity.
Too little adhesion	Excessive moisture	Check liquid levels in formulation. Add starch. Allow product to cool completely before finishing.
Too soft	Excessive moisture	Check liquid levels in formulation. Add sucrose.
Icing sticks to wrapper or packaging	Too hygroscopic	Add starch. Decrease water content. Cool product completely before packaging.
Fruit fillings thin, runny	Thickening agent hydrolyzes at low pH or breaks down with heat	Increase pH; decrease acid. Use crosslinked starch for thickening agent
Fruit fillings exhibit syneresis and poor freeze-thaw stability	Moisture migration	Use substituted (stabilized) starches.
Poor or off-flavor in cream fillings	Thickening agent masks delicate flavors	Use clean-flavor starches such as tapioca.

EXTRUDED PRODUCTS

Symptom	Possible Causes	Changes to Consider
Product lacks crispness	Poor, weak expansion framework	Increase amylose content if high-shear conditions are used. Increase amylopectin content if low-shear conditions are used.

Poor cutting, shape	Low dough viscosity or strength	Increase amylose content if high-shear conditions are used. Increase amylopectin content if low-shear conditions are used. Adjust moisture content.
Sticky dough after extrusion	Formation of dextrins and short chain polymers during processing	Increase amylose content. Add more highly crosslinked starches. Adjust moisture content. Decrease shear during processing.
Low dough viscosity	Low dough strength	Increase amylopectin content or add crosslinked starches for high-shear conditions. Adjust moisture content.
Nonuniform sheet thickness from rolling extruded ropes	High water absorption by starch	Decrease water content. Choose starch with low water-holding capacity.
Product sticks to teeth; "melts" easily in mouth	Formation of dextrins and short chain polymers during processing	Increase amylose content. Add more highly crosslinked starches.

BATTERS AND BREADINGS

Symptom	Possible Causes	Changes to Consider
Low viscosity; poor adhesion	Low level of thickening agent	Increase starch level. Check solids level.
Nonuniform coating	Poor dispersion of ingredients in formula	Blend well.
	Spices, solids settle out of batter formula	Add starch to increase viscosity. Maintain agitation.
Coating falls off after final processing (frying, freezing, etc.)	Poor binding ability	Add converted starch.
Finished product oily or greasy	High level of oil pick-up and retention in product	Add high-amylose starch.
Coated product too dry	Poor moisture retention	Avoid overfrying. Select starch with better water-holding capacity.
Coating soggy or not crisp	Poor structure formation	Lower water content. Add high-amylose starch.

References

1. Abboud, A. M., and Hoseney, R. C. 1984. Differential scanning calorimetry of sugar cookies and cookie doughs. Cereal Chem. 61:34-37.
2. Kulp, K., and Loewe, R., Eds. 1990. *Batters and Breadings in Food Processing*. American Association of Cereal Chemists, St. Paul, MN.

CHAPTER 8

Sauces, Dressings, and Other Applications

Sauces, Gravies, and Soups

This category represents a wide variety of food products with diverse demands on starch functionality. Starch functions not only as a thickener, but also provides the texture and mouthfeel to these products. Chemically modified starches have allowed the commercial production and distribution of soups, sauces, and gravies. The high-heat processing and long shelf life usually associated with these products demand the use of highly stable starches.

The selection of a starch depends on the pH of the system and the production process, including cooking temperature and shear, that the starch will undergo. These parameters determine the amount of crosslinking required, and the storage conditions of the product dictate the substitution level required. A substituted starch maintains freeze-thaw stability and long-term water-holding capacity in products stored at ambient or refrigeration temperatures. Improper selection of a starch may reduce the shelf life and result in an inferior product.

Soups, sauces, and gravies typically fall into two categories, low acid (pH > 4.5) and acid. Low-acid products include chicken and turkey gravy, dairy-based pasta sauces, and cheese sauces. Acid products include barbecue and chili sauces, tomato-based pasta sauces, and salsa. Acidity level dictates to a large extent how the product will be processed. Low-acid foods must undergo a rigorous heat process to assure microbiological stability (see Chapter 5).

In This Chapter:

RETORTED PRODUCTS

Low-acid foods can be preserved via a *retort ("canning") process*, and various types of modified starches are available for use either alone or in combination in such a system. A retort is essentially a commercial-scale pressure cooker that is used in a canning operation to thermally process food at high temperatures (e.g., 121°C [250°F]) and increased pressures for various lengths of time. Each retorted food product has its own thermal processing requirements, which are under strict control. The food container is usually a metal can, glass jar, or retort-stable plastic package.

One type of starch used for a retort process is the so-called "fill viscosity" starch. As the name implies, a fill viscosity starch gelatinizes easily prior to retorting to give viscosity during the initial mixing,

Retort ("canning") process—A thermal process in which high temperatures and pressure are used to preserve foods for relatively long-term storage typically under ambient conditions.

cooking, and filling stages of the canning operation. Adequate viscosity during these steps is critical for keeping food particles homogeneously suspended in the kettle or mixer, thus allowing for can-to-can consistency during filling. Pre-retort viscosity also helps to prevent splattering of the product during the filling step. A fill viscosity starch functions during the pre-retort phase but is usually not stable during retorting. In fact, many fill viscosity starches are designed to break down to a watery consistency during the retort process (Fig. 8-1).

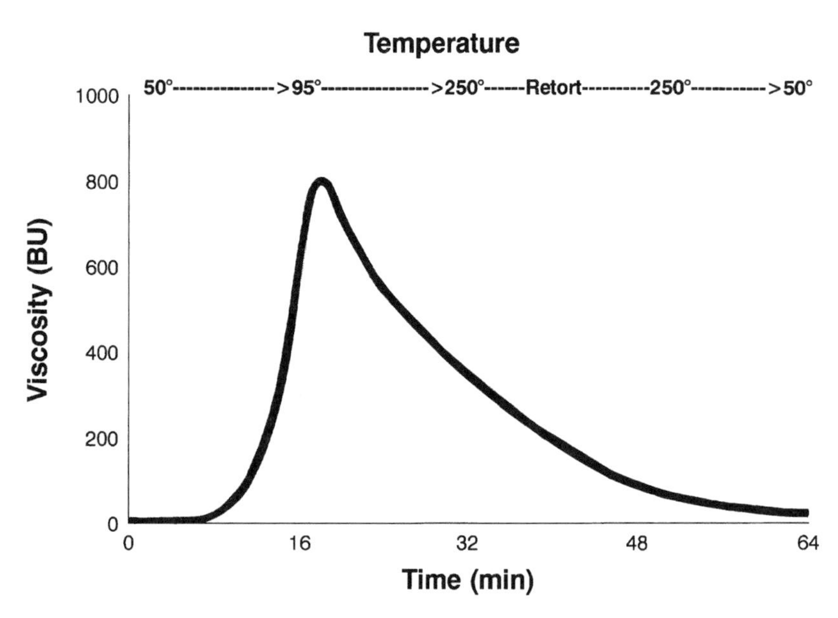

Fig. 8-1. Typical viscosity profile for a fill viscosity starch.

The other common type of modified starch that is typically used in canning operations is a viscosifying starch. This type of modified starch is almost always crosslinked (and usually substituted), thus allowing for relatively consistent viscosity during retorting. These starches typically dictate final product viscosity.

Canning starches that have a very high level of crosslinking can provide a "thin-thick" type of viscosity profile during retorting (Fig. 8-2). In this case, the viscosity or thickening does not develop until the system

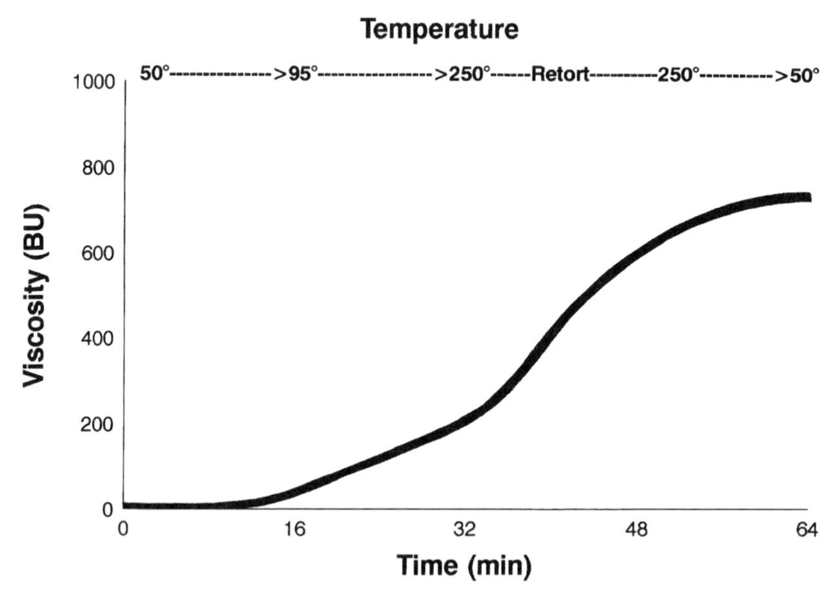

Fig. 8-2. Typical viscosity profile for a heat-penetration starch.

has reached the high temperatures associated with a retort process. This type of starch may allow for improved heat penetration during retorting and hence more efficient commercial sterilization. With this high degree of modification, the type of starch base is not as critical. By acting sequentially, fill viscosity starches, viscosifying starches, and thin-thick (heat penetration) starches are often used in combination to provide the desired properties in the final product.

Acidic food products (pH ≤ 4.5) are usually thermally processed at lower temperatures (e.g., 93°C [200°F]). These so-called "hot-fill" operations are typically used for tomato and other fruit-based products, which are inherently acidic. In these types of systems, it is not necessarily the thermal process that is of concern with respect to starch performance, but rather the effect of the acid on starch polymer hydrolysis and viscosity breakdown. In order to withstand an acid environment, a moderately to highly crosslinked starch is typically required.

Other shelf-stable applications that are thermally processed include stews, vegetables (e.g., "cream-style" corn), baby foods, pet foods, and fillings. Depending upon the pH of the particular system, these applications are also processed as either low-acid or acidic foods.

Many starches that are used in the canning industry are declared *thermophile*-free (TF grade) (1). Before a starch can be labeled and marketed as TF by a starch manufacturer, it must conform to microbial specifications set forth by the National Food Processors Association. The behavior of TF grade starch in terms of viscosity and stability is similar to that of its non-TF counterparts.

FROZEN AND DRY PRODUCTS

Some soups and sauces are first cooked and then frozen. These frozen "heat-n-serve" products require starches that not only withstand the initial processing requirements, but also provide freeze-thaw stability during distribution and storage. In these particular applications, starches that are both crosslinked and moderately to highly substituted are a must.

Starches, both cook-up and pregelatinized varieties, are typically used at a level of 2–5% to provide viscosity and a smooth, creamy texture to gravies and sauces made from dried mixes. The starch should have a low moisture content to facilitate the free flow of the mix and should provide good storage stability so that the prepared product will not gel or lump when cooled. Specific starches are also available that provide good uniformity and dispersion to microwavable mixes.

Dressings

As defined in Section 169.150 of the Code of Federal Regulations (Title 21), "salad dressing" is the emulsified, semisolid food prepared from vegetable oil(s), an acidifying ingredient (e.g., vinegar), an egg

Thermophile—Microorganism that can grow at temperatures above 45–50°C.

yolk-containing ingredient, and a starchy paste. Optional ingredients to these high-acid, high-fat products may include salt, sweeteners, spices, sequestrants, and stabilizers or thickeners. In order to be called "salad" dressing, the product cannot contain less than 30% vegetable oil (by weight). "Dressing," however, is quite popular (particularly the no- and low-fat varieties) and ranges from 0 to about 30% fat.

Salad dressing is a challenging application for starch and requires the proper selection of starch to provide a stable product. Salad dressing is a highly acidic system, with a pH as low as 2.8 and a total titratable acidity of 2.5% in accordance with the standard of identity. These dressings are usually heat processed and are then subjected to high shear, which produces a fine emulsion. In addition to these manufacturing stresses, a salad dressing may have a 12-month shelf life. A salad dressing starch must have the proper amount of chemical modification to withstand the manufacturing process and to provide an adequate shelf life.

Starches designed for spoonable salad dressings are usually a combination of *common starch* (e.g., dent corn starch) and waxy starch (e.g., waxy corn starch) that have been chemically modified to varying degrees and blended in differing ratios to give the desired final product characteristics. The amylose-containing starch provides the "set" or spoonable nature to the dressing. Waxy starch provides viscosity and stability and is generally moderately to highly crosslinked and substituted. When waxy and common starches are blended, the spoonable nature is achieved.

Pourable salad dressings are typically produced by a cold process, which means that there is no heating or cooking step in the manufacture of these products. Granular pregelatinized starches provide the desired stability and quality attributes to these dressings and can be highly modified, common, or waxy starches. Starch shortens the texture and adds creaminess and richness to a pourable salad dressing compared with a dressing thickened with xanthan gum, for example. However, it is common to blend xanthan and starch as stabilizers in these dressings.

EMULSION STABILITY

In addition to producing a tasty and flavorful product, a key objective of the salad dressing manufacturer is to generate a stable emulsion with good final texture. A stable emulsion is a function of the ingredients that are present and the process for creating the emulsion. An ingredient such as egg yolk, which contains the emulsifier *lecithin*, helps to reduce surface tension and prevent oil coalescence. Other emulsifiers can be added to retard the migration of oil droplets to the surface (a process called "creaming") by promoting the smaller droplets that slow the separation of oil and water.

Starch plays a key role in emulsion stability by physically separating the oil droplets. In order to function adequately, the starch must be able to tolerate the high-shear process used to generate the emul-

Common starch—Any non-waxy amylose-containing starch; however, the term is often used to describe dent corn starch.

Lecithin—Trivial name for the emulsifier phosphatidylcholine, a phospholipid usually isolated from soybean or egg yolk.

sion. Processing equipment plays an indisputable role in creating (and indirectly stabilizing) an emulsion. Salad dressings are typically prepared by using a high-shear mixer or homogenizer such as a colloid mill, which creates a homogenous mixture of ingredients and an even dispersion of the oil. Because of this high shear, the oil is dispersed in tiny droplets, typically less than 1 μm in diameter, that are evenly distributed throughout the water-continuous phase of the product. The even distribution of tiny oil droplets is critical for a stable emulsion.

TEXTURE

In addition to emulsion stabilization, starch plays a critical role in defining the texture of the dressing. Native starch by itself does not provide adequate viscosity and texture, since it cannot withstand the severe processing conditions (i.e., high shear) and acid environment (i.e., pH 3–4) associated with salad dressing. Products prepared with only native waxy corn starch, for example, are typically described as being cohesive or gummy. Dressings prepared with unmodified amylose-containing starch, such as dent corn starch, have a tendency to form a gel after processing. Although this may be somewhat desirable in a dressing, a chemically modified starch such as a moderately to highly crosslinked and substituted product is usually a requirement for salad dressing manufacture. In some cases, both native and modified starches are used in combination to give the desired short, smooth texture. A typical usage level for starch in a salad dressing is 2–6% (by weight) of the total formula. Given the history of salad dressing production and the business volume that it represents, most starch companies have products designed specifically for particular salad dressing applications.

Meat Products

The use of modified starch in meat products is regulated by the U.S. Department of Agriculture's Food Safety and Inspection Service and is noted in Part 318.7 of the Code of Federal Regulations (Title 9). Given the complexity of the issues surrounding the use of such substances in meat products as well as the large number of processed meat, poultry, and seafood products that are currently being marketed, the focus here is on only the functional characteristics of starch in meat-based products and not on its regulatory status. In order to fully appreciate the role of starch as a binder or extender in processed meats, a brief review of some important characteristics of meat products is merited.

Meat such as beef, pork, and lamb is generally categorized as "red" meat; chicken, turkey, and duck as poultry meat; and fish and shellfish as seafood (2). The composition of typical mammalian meat muscle is shown in Table 8-1 (3). Of the constituents listed, water represents the majority of lean meat tissue mass. With this in mind, it is easy to understand why water-holding capacity is a key measurement

TABLE 8-1. Approximate Composition of Typical Mammalian Muscle[a]

Component	Wet Weight (%)
Water	75.0
Protein	19.0
Lipid	2.5
Carbohydrate	1.2
Miscellaneous (e.g., inorganic elements)	2.3

[a] Adapted from (3).

of processed meat products and why moisture control during and after meat processing is critical to the quality of the final product.

COMMINUTED PRODUCTS

The *comminution* of meat and its reformation into products such as frankfurters and bologna represent a large segment of the meat industry. The degree of comminution can vary widely from coarse chunks to very small particles, which are akin to a viscous paste that is essentially an emulsion. Frankfurters and bologna are examples of emulsified meat products. As in any emulsion, various factors, e.g., fat particle size, temperature during emulsification, pH, viscosity, and salt content, affect the formation and stability of a meat emulsion.

Starch, particularly modified cook-up starch, is used at 1–3% and is well-suited for a wide variety of applications in the meat industry. It has found application in emulsified meat products such as bologna and frankfurters as well as in meat- and poultry-based rolls and loaves. Modified starches most often used in meat products are those that have been both lightly or moderately crosslinked and moderately or highly substituted. Starches with high viscosity profiles and high water-holding capacities are typically good first choices when it comes to improving the texture and shelf life of processed meat products. Parameters to monitor in the product include water-holding capacity, degree of purge, cuttability or sliceability, and freeze-thaw stability.

Comminution—Mechanical subdivision of meat (usually less expensive cuts) into small pieces, which are then restructured or reformed into various meat products.

Surimi—Japanese word referring to imitation seafood (e.g., crab or lobster) produced by mincing and reforming fish meat along with added flavors.

SURIMI

Starch is an important ingredient in *surimi* and is used as a filler and for its water-holding capacity. It also has a significant effect on the texture and physical characteristics of the surimi gel. Starch blends are used to improve the moistness of the gel. Modified starches reduce elasticity and firmness but increase the freeze-thaw stability of the product. The starches chosen depend on the end use of the gel. In cold applications, blends of native and modified amylose-containing starches provide better texture. In hot applications, waxy starch is preferred in the blend (4).

Troubleshooting

This troubleshooting section lists some common problems, causes related to starches, and suggestions for changes to consider in the formulation and processing of soups, sauces, and gravies, dressings, and processed meat products. However, a problem is often the result of a combination of factors that have to be considered: the amount and type of starch used and the kind of modification it has undergone, processing conditions (e.g., heat, shear, moisture, and time), effects of other ingredients such as fat and sugar, and factors such as pH and storage conditions. The solution may involve any one or more of these factors. This guide may serve as a starting point in looking at specific products.

SOUPS, SAUCES, AND GRAVIES

Symptom	Possible Causes	Changes to Consider
Syneresis; freeze-thaw instability	Moisture migration; starch degradation	Add substituted (stabilized) starch
Low viscosity	Inadequate starch swelling, gelatinization	Increase starch content. Decrease water content. Check cooking time and/or temperature.
	Starch degradation (e.g., over-shearing, enzyme activity)	Adjust pH. Check starch selection.
Too thick, viscous	Too much starch swelling or gelatinization	Decrease starch content. Increase water content. Check cooking time and/or temperature.
Poor or off-flavors	Poor quality of incoming ingredients	Check quality of incoming ingredients.
	Off-notes from starch	Use tapioca (clean-flavor) starch.
	Undercooked starch	Check cooking time and/or temperature.
Lumpy product	Poor dry ingredient dispersibility	Check moisture content of starch. Add maltodextrin for preblending. Increase mixing.

DRESSINGS

Symptom	Possible Causes	Changes to Consider
Gummy texture	Too much amylopectin or gums	Decrease waxy starch and/or gum content.
	Overcooking	Decrease cooking time and/or temperature.
Low viscosity	Inadequate starch swelling	Increase level of thickening agents.
	Starch degradation	Add crosslinked and substituted starch. Check cooking time and/or temperature.
Texture of pourable dressing too short	Too much amylose present	Increase waxy starch content.
Spoonable dressing too viscous, short	Viscosity level provided by starch too high	Check starch selection. Decrease starch content.

Texture of spoonable dressing too long; dressing is pourable	Too much amylopectin present	Check starch selection. Check starch swelling.
	Overcooking	Check cooking time and/or temperature to prevent overcooking.
MEAT PRODUCTS		
Symptom	**Possible Causes**	**Changes to Consider**
Poor water-holding capacity; low freeze-thaw stability	Lack of water-binding ingredients	Add substituted (stabilized) starch. Select a starch with higher water-holding capacity.
Poor bite; soft texture	Structure not developed	Add substituted (stabilized) starch. Check starch selection.

References

1. Brock, T. D., Smith, D. W., and Madigan, M. T. 1984. The microbe and its environment. Pages 235-270 in: *Biology of Microorganisms*, 4th ed. Prentice-Hall, Englewood Cliffs, NJ.
2. Forrest, J. C., Aberle, E. D., Hedrick, H. B., Judge, M. D., and Merkel, R. A. 1975. *Principles of Meat Science.* W. H. Freeman and Company, San Francisco.
3. Lawrie, R. A. 1985. *Meat Science,* 4th ed. Pergamon Press, New York.
4. Mauro, D. J. 1996. An update on starch. Cereal Foods World 41:776-780.

CHAPTER 9

Special Topics

In This Chapter:

- Fat Replacement
- Emulsion Stabilizers
- Encapsulating Agents
- Resistant Starch

Fat Replacement

If one spends any time in a modern supermarket, it is clear that there are a large number and a wide variety of food products in the reduced-, low-, and no-fat categories. These products are in all areas of the supermarket, represent basically all types of food, and are made for all kinds of distribution systems. The reduced-fat trend is growing and is not likely to change given today's interest in health and well-being.

Fat serves many functions in foods. It contributes to the appearance, texture, flavor, and mouthfeel of products in many ways. Therefore, replacing fat is not easy, and a great deal of effort has been spent identifying ways to do so. As the number of reduced-fat products continues to rise, so has the number of ingredients used in the formulation of reduced-fat foods. Generally, the fat-replacing ingredients used today can be categorized as carbohydrate-based substitutes, tailored fat-replacing compounds, calorie-reduced fats, protein-based substitutes, emulsifiers, and combination products (1,2). This discussion is limited to starch-based substitutes.

Starch-based fat replacers come in many varieties from a large number of suppliers. It is estimated that there are now about 40 starch-based products used in fat-replaced formulations (3). The value of starch as a fat-replacing ingredient is its inherent energy value of 4 kilocalories per gram compared with 9 kilocalories per gram of fat. If, for example, a starch-based *fat mimetic* consists of a gel containing about 25% starch and 75% water, the caloric value is about 1 kilocalorie per gram, resulting in an effective fat reduction of about 90%.

Starch-based fat-replacement technology began with patents on potato starch hydrolysis products (4,5). These patents claim thermoreversible gels in aqueous food systems that have a texture and mouthfeel approximating fat in applications such as dressings and frozen desserts.

Starches can be manipulated to yield gels that impart textural and mouthfeel properties that approximate those of fat. The particle size of the starch granules, dispersibility of starch granule components, molecular weight of the starch polymers, and water-binding ability are controlled to yield the fat-replacing qualities required. Starch-based fat-replacing ingredients can be either starches or converted

Fat mimetic—Fat substitute that mimics the properties of fat.

starch products such as maltodextrins. The maltodextrins used as fat replacers are typically low-DE (high molecular weight) products.

Applications of starch-based fat replacers are numerous. The primary applications are in baked products, dressings, dairy products, sauces, dips, and mayonnaise-like products. Suppliers should be contacted for specific information concerning product applications and formulation details.

Emulsion Stabilizers

Starches are naturally hydrophilic and bind large amounts of water, especially in the gelatinized form. They can be modified, however, so that they contain hydrophobic regions as well. Because of this dual property, these types of modified starches are able to interact well with both water and oil and thus have a stabilizing effect in oil-water mixtures. The starches used for this purpose are generally made by reacting starch or dextrin with substituted dicarboxylic acids or anhydrides, usually substituted succinic anhydride (6). The resultant products are called starch alkenyl succinates (see Chapter 4).

A distinction should be drawn between the effects of these types of starch derivatives in oil-water mixtures and those of conventional emulsifiers. The key point is that these types of starches are not true emulsifiers but rather emulsion stabilizers. They must be either pregelatinized or heated to ensure that they disperse well enough in the water phase to have a stabilizing effect at the oil-water interface. They can also have a significant effect on the viscosity of the emulsion and can be manipulated by varying molecular weight and other properties to yield many desired viscosities.

The molecular weight of starch can be lowered to the desirable range by acid degradation (see the section on starch-conversion methods in Chapter 4), thus producing starch alkenyl succinates with reduced viscosity. 1-Octenylsuccinic anhydride (OSA)-treated starch is FDA approved at a maximum allowable level of 3% in a formulation. Stabilized emulsions can be of low viscosity or have a very thick, creamy consistency. A major application of these starches is stabilizing nonalcoholic beverage emulsions such as soft drink concentrates, especially orange oil concentrates. There is also a large variety of salad dressings and sauces formulated with OSA-treated starches.

Encapsulating Agents

An encapsulation process is often employed to protect an ingredient or product until it is used by the consumer. Some ingredients within a food product are encapsulated to delay functionality during processing or to preserve their functionality during distribution and storage. Ingredients including leavening agents, vitamin enrichments, aromas, flavors, nuts, and fruits and finished food products such as breakfast cereals can be encapsulated. Encapsulating agents can be either fats, proteins, or carbohydrates or mixtures and derivatives thereof.

Starch-derived dextrins have been used as encapsulating agents for spray-dried flavors for many years. One of the first successful methods employing this technology was a patented process that involved the use of dextrins derived from oxidized starches (7). Prior to the development of this process, a major drawback to the use of starch-based encapsulating agents was their susceptibility to Maillard browning during the encapsulation procedure, resulting in unacceptable flavors, aroma, and colors. The use of oxidation and other starch-modification techniques has eliminated these problems (8).

OSA-treated starches are also used routinely to encapsulate flavors (6). Generally, a low-viscosity, OSA-treated starch is employed in a spray-drying operation. Upon drying, OSA-treated starch encapsulates usually have better flavor retention than those encapsulated with gum arabic or conventional dextrins. These encapsulates are easily hydrated, readily combine with water, and quickly release the flavoring agent they carry.

There are cases in which it is desirable that the encapsulate be water repellent and/or that the release of the flavor be more gradual. This has been accomplished with a technology patented by Wurzburg and Cole (9). OSA-treated starch is reacted with a polyvalent metal ion (e.g., calcium or magnesium), which renders the encapsulate more hydrophobic. Temperature-sensitive materials, such as vitamins, are encapsulated with high-amylose, OSA-treated starch (10). These encapsulates set up quickly and do not require high temperatures to dry.

Resistant Starch

The term "resistant starch" (RS) was adopted by Hans Englyst, a British physiologist, in the early 1980s. RS is defined as dietary starch that does not digest in the small intestine (11,12). RS, therefore, behaves like dietary fiber and may have potential as a health-related ingredient in foodstuffs. The primary difference between digestible and resistant starches is the accessibility of the starch to digestive processes and subsequently the ease with which the glycosidic bonds contained within the starch molecules can be severed (13). Starch that is protected by cell walls or other barriers to the actions of the digestive enzymes and acidic conditions of the digestive tract is considered Type I RS. Examples of this type of RS would be the starch in partially milled seeds and grains. Highly crystalline native starches such as those from raw bananas or potatoes cannot be reduced to degradation products that can be absorbed through the small intestine and are considered Type II RS. If the starch in a food system is gelatinized, subsequently cooled, and allowed to retrograde, a similarly resistant starch is created (Type III). Corn flakes, cooked potatoes, and canned peas contain some retrograded starch that is resistant. Finally, it is

TABLE 9-1. Classification of Resistant Starch

Category	Definition
Type I	Physically inaccessible starch that is locked in plant material such as milled grains, seeds, and legumes
Type II	Native granular starch found in uncooked food such as bananas
Type III	Indigestible starch that forms after heat and moisture treatment; may be present in foods such as cooked potatoes
Type IV	Resistant starch that is produced by chemical or thermal modification

also possible to make starch resistant by modifying it and altering its native structure (Type IV).

Table 9-1 summarizes the classification of RS into these four types based upon resistance to digestion, granular structure, relationship to retrogradation, and the type of starch modification. In Type IV RS, starch modifications that result in glycosidic linkages other than α-1,4 and α-1,6 substitutions and cross-linkages impart increased resistance to amylases (14). Processes for preparing commercial forms of RS generally involve the recovery of highly retrograded, crystalline regions from starch, particularly high-amylose starch (15,16). The RS portion of the starch processed by these methods generally ranges from about 10 to 40%.

Because RS assays as total dietary fiber, it is considered a functional fiber similar to more common fibers such as oat bran. Commercial forms of RS have been proposed as ingredients that can be used for fiber fortification of foods such as breads, snacks, and breakfast cereals.

Possible beneficial effects of using resistant starch include lowered pH of the colon, formation of short-chain fatty acids in the colon, increased fecal bulk, protection against colon cancer, improved glucose tolerance, and lowered blood lipid levels.

The health and nutritional implications of RS are potentially similar to those of nonstarch polysaccharides such as cellulose, hemicelluloses, pentosans, and β-glucans, which comprise the cell walls of plants. Therefore, some of the beneficial effects attributed to dietary fiber might also be present with RS (17). These proposed responses, considered either alone or in combination, contribute to the purported health-related benefits of RS. For example, a lower pH in the colon sets up an environment that favors the growth of lactobacilli, microorganisms considered beneficial to the gastrointestinal tract. Short-chain fatty acids (e.g., butyric acid) may help intestinal cell functionality and aid in the prevention of colon cancer. Overall, the role of RS in nutrition and diet appears to be quite significant. The physiological changes that occur with the ingestion of RS continue to be an area of tremendous scientific interest.

References

1. Summerkamp, B., and Hesser, M. 1990. Fat substitute update. Food Technol. 44:92, 94, 97.
2. Haumann, B. F. 1992. Here's a list of who's producing what. INFORM 3:1284-1287.
3. Alexander, R. J. 1995. Fat replacers based on starch. Cereal Foods World 40:366-368.
4. Richter, M., Schierbaum, F., Augustat, S., and Knock, K. D. 1976. Method of producing starch hydrolysis products for use as a food additive. U.S. patent 3,962,465.

5. Richter, M., Schierbaum, F., Augustat, S., and Knock, K. D. 1976. Method of producing starch hydrolysis products for use as a food additive. U.S. patent 3,986,890.
6. Trubiano, P. C. 1986. Succinate and substituted succinate derivatives of starch. Pages 131-147 in: *Modified Starches: Properties and Uses*. O. B. Wurzburg, Ed. CRC Press, Boca Raton, FL.
7. Evans, R. B., and Herbst, W. 1964. Method for using dextrin as an encapsulating agent. U.S. patent 3,159,585.
8. Daniels, R. 1973. Encapsulation processes. Pages 297-300 in: *Edible Coatings and Soluble Packaging*. R. Daniels, Ed. Noyes Data Corporation, Park Ridge, NJ.
9. Wurzburg, O. B., and Cole, H. M. 1963. U.S. patent 3,091,567.
10. Wurzburg, O. B., and Trubiano, P. C. 1970. U.S. patent 3,499,962.
11. Saura-Calixto, F., Goni, I., Bravo, L., and Manas, E. 1993. Resistant starch in foods: Modified method for dietary fiber residues. J. Food Sci. 58:642-643.
12. European Flair—Concerted Action on Resistant Starch (EURESTA). 1991. Physiological implications of the consumption of resistant starch in man. Flair Concerted Action No. 11, Newsl. III.
13. Stephen, A. M. 1995. Resistant starch. Pages 453-458 in: *Dietary Fiber in Health and Disease*. D. Kritchevsky and C. Bonfield, Eds. Eagan Press, St. Paul, MN.
14. Eerlingen, R. C., and Delcour, J. A. 1995. Formation, analysis, structure and properties of type III enzyme resistant starch. J. Cereal Sci. 22:129-138.
15. Iyengar, R., Zaks, A., and Gross, A. 1991. Starch-derived, food-grade, insoluble bulking agent. U.S. patent 5,051,271.
16. Chiu, C.-W., Henley, M., and Altieri, P. 1994. Process for making amylase resistant starch from high amylose starch. U.S. patent 5,281,276.
17. Bjorck, I., and Asp, N.-G. 1994. Controlling the nutritional properties of starch in foods—A challenge to the food industry. Trends Food Sci. Technol. 5:213-218.

Glossary

Amylase—Any one of several starch-degrading enzymes common to animals, plants, and microorganisms.

α-Amylase (α-1,4-glucan-4-glucanohydrolase)—Enzyme that acts to degrade starch polymers at an internal site anywhere in an amylose or amylopectin molecule by hydrolyzing α-1,4 linkages.

β-Amylase (α-1,4-glucan maltohydrolase)—Enzyme that cleaves alternate glycosidic bonds of starch in α-1,4 chains in a stepwise fashion starting at the nonreducing end.

Amylopectin—A very large, branched, D-glucopyranose polymer of starch containing both α-1,4 and α-1,6 linkages. The α-1,6 linkage represents the bond at the polymeric branch point.

Amyloplasts—Organelles in plant cells that synthesize starch polymers in the form of granules.

Amylose—An essentially linear polymer of starch composed of α-1,4-linked D-glucopyranose molecules. A small number of α-1,6-linked branches may be present.

Ash—Mineral and salt fraction, typically calculated by determining the amount of residue remaining after incineration of a sample.

Birefringence—Phenomenon that occurs when polarized light interacts with a highly ordered structure, such as a crystal. A crossed diffraction pattern, often referred to as a "Maltese cross," is created by the rotation of polarized light by a crystalline or highly ordered region, such as that found in starch granules.

Casein—A milk protein that is a complex of protein and salt ions, principally calcium and phosphorus, present in the form of large, colloidal particles called micelles.

Cellulose—The most abundant carbohydrate polymer on earth, found only in plants, comprising β-1,4-linked D-glucopyranose units.

Clathrate—An inclusion complex wherein a "host" molecule entraps a second molecular species as the "guest."

Comminution—Mechanical subdivision of meat (usually less expensive cuts) into small pieces, which are then restructured or reformed into various meat products.

Common starch—Any non-waxy amylose-containing starch; however, the term is often used to describe dent corn starch.

Compound starch granules—Naturally occurring aggregates of individual starch granules.

Cook-up starch—Any granular starch that requires water and heat in order to gelatinize and paste.

Corn starch—Common corn starch composed of approximately 25% amylose and 75% amylopectin.

Crosslinking—Chemical modification of starch that results in covalently bonded inter- and intramolecular bridges between starch polymers.

Cyclodextrin—A circular molecule of α-1,4-linked glucose units.

Cyclodextrin glycosyltransferase—Enzyme that cleaves starch and cyclizes the resulting glucose chain into a cyclodextrin ring.

Damaged starch—Starch that has been mechanically disrupted by shearing during processing.

Debranching enzyme—Enzyme, such as isoamylase and pullulanase, that hydrolyzes the α-1,6 linkages of starch.

Degree of polymerization (DP)—The molecular size of a polymer. In this case, it refers to the number of α-1,4-linked D-glucopyranose units in a starch chain.

Degree of substitution (DS)—Measurement of the average number of hydroxyl groups on each D-glucopyranosyl unit (commonly called an anhydroglucose unit [AGU]) that are derivatized by substituent groups. Since the majority of AGUs in starch have three hydroxyl groups available for substitution, the maximum possible DS is 3.

Dextrose equivalent (DE)—Indication of the reducing sugar content calculated as the percent anhydrous dextrose of the total dry substance. Pure dextrose has a DE of 100.

Endoenzyme—Enzyme that splits bonds anywhere along a polymer chain.

Endosperm—Interior portion of the wheat kernel containing the gluten and starch, which comprise flour; the fuel source for a sprouting wheat plant.

Exoenzyme—Enzyme that splits only terminal bonds in a polymer chain.

Fat mimetic—Fat substitute that mimics the properties of fat.

Freeze-thaw (F-T) stability—Ability of a product to withstand cold temperature cycling and/or prolonged storage at reduced temperatures.

Gel—A liquid system that has the properties of a solid.

Gelatinization—Collapse (disruption) of molecular orders within the starch granule manifested by irreversible changes in properties such as granular swelling, native crystalline melting, loss of birefringence, and starch solubilization.

Gelatinization temperature—A narrow temperature range at which starch granules begin to swell, lose crystallinity, and viscosify the cooking medium.

Germ—Embryo of the wheat plant within the kernel.

Glucoamylase (α-1,4-glucan glucohydrolase)—Enzyme that removes the glucose units consecutively from the nonreducing ends of starch polymers by hydrolyzing both α-1,4 and α-1,6 linkages.

D-Glucopyranose—The ring form of the monosaccharide D-glucose.

Glycosidic bond—Covalent linkage formed between D-glucopyranose units.

Hardness—Amount of force required to crush a kernel of wheat. Hard wheats require more energy to mill to flour and generate more damaged starch during the process than soft wheats.

High-amylose corn starch—Starch isolated from a hybrid corn plant that contains greater than about 40% amylose. Some high-amylose corn starches now contain as much as 90% amylose.

Hydrophilic—Attracted to water or polar regions of molecules.

Hydrophobic—"Water-fearing." The term is usually used to describe nonpolar substances, e.g., fats and oils, that have little or no affinity for water.

Lecithin—Trivial name for the emulsifier phosphatidylcholine, a phospholipid usually isolated from soybean or egg yolk.

β-Limit dextrin—Product resulting from the action of β-amylase on a branched starch polymer such as amylopectin.

Molar substitution (MS)—Level of substitution in terms of moles of monomeric units (in the polymeric substituent) per mole of AGU. The MS can be greater than 3, given the substituent's ability to react further.

Mono- and **Diglycerides**—Glycerol molecules with one or two fatty acids attached, respectively.

Native starch—Any granular starch that has been isolated from the original plant source but has not undergone subsequent modification; i.e., unmodified starch.

Paste—Starch in which a majority of the granules have undergone gelatinization, giving it a viscosity-forming ability. Pasting involves granular swelling and exudation of the granular molecular components.

Peak viscosity—The point at which, during heating in water, gelatinized starch reaches its maximum viscosity.

Reducing sugar—A monosaccharide, disaccharide, oligosaccharide, or related product capable of reducing an oxidizing ion. A common test for the measurement of reducing sugars involves the reduction of cupric ions (Cu^{+2}) to cuprous ions (Cu^{+}).

Retort ("canning") process—A thermal process in which high temperatures and pressure are used to preserve foods for relatively long-term storage typically under ambient conditions.

Retrogradation—Process during which starch chains begin to reassociate in an ordered structure. Two or more starch chains initially form a simple juncture point, which then may develop into more extensively ordered regions and ultimately, under favorable conditions, to a crystalline order.

Set-back—Generally, the reassociation of solubilized starch polymers and insoluble granular fragments during the cooling phase of a viscoamylograph analysis.

Spherosomes—Fat-containing organelles in endosperm cells.

Starch granules—Naturally occurring, partially crystalline, discrete aggregates of amylose and amylopectin.

Substitution—Chemical modification of starch resulting in the addition of a chemical blocking group between starch polymers and involving derivatization with a monofunctional reagent through ester or ether formation.

Surimi—Japanese word referring to imitation seafood (e.g., crab or lobster) produced by mincing and reforming fish meat along with added flavors.

Swelling power—Swollen sediment weight (in grams) per gram of dry starch.

Syneresis—The separation of a liquid from a gel; weeping.

Thermophile—Microorganism that can grow at temperatures above 45–50°C.

Thin-boiling starch—Acid-hydrolyzed starch used to reduce the hot viscosity of a paste so that higher concentrations of starch can be dispersed without excessive thickening.

Triglyceride—Three fatty acids attached to a glycerol molecule.

Viscoelastic paste—A paste that possesses properties of both a viscous (liquid-like) and an elastic (solid-like) substance.

Water fluidity (WF)—Scale used to determine the degree of starch conversion.

Waxy starch—"Amylopectin" starch from certain cereals such as corn and rice. The amylopectin content of waxy starches is generally about 95% or more. The term "waxy" has nothing to do with the presence of wax, but rather describes the appearance of the cross section of the mutant corn kernel from which it was first isolated.

X-ray pattern—Pattern obtained when a crystal is irradiated with X rays. This pattern is distinctive to the crystal structure.

Index